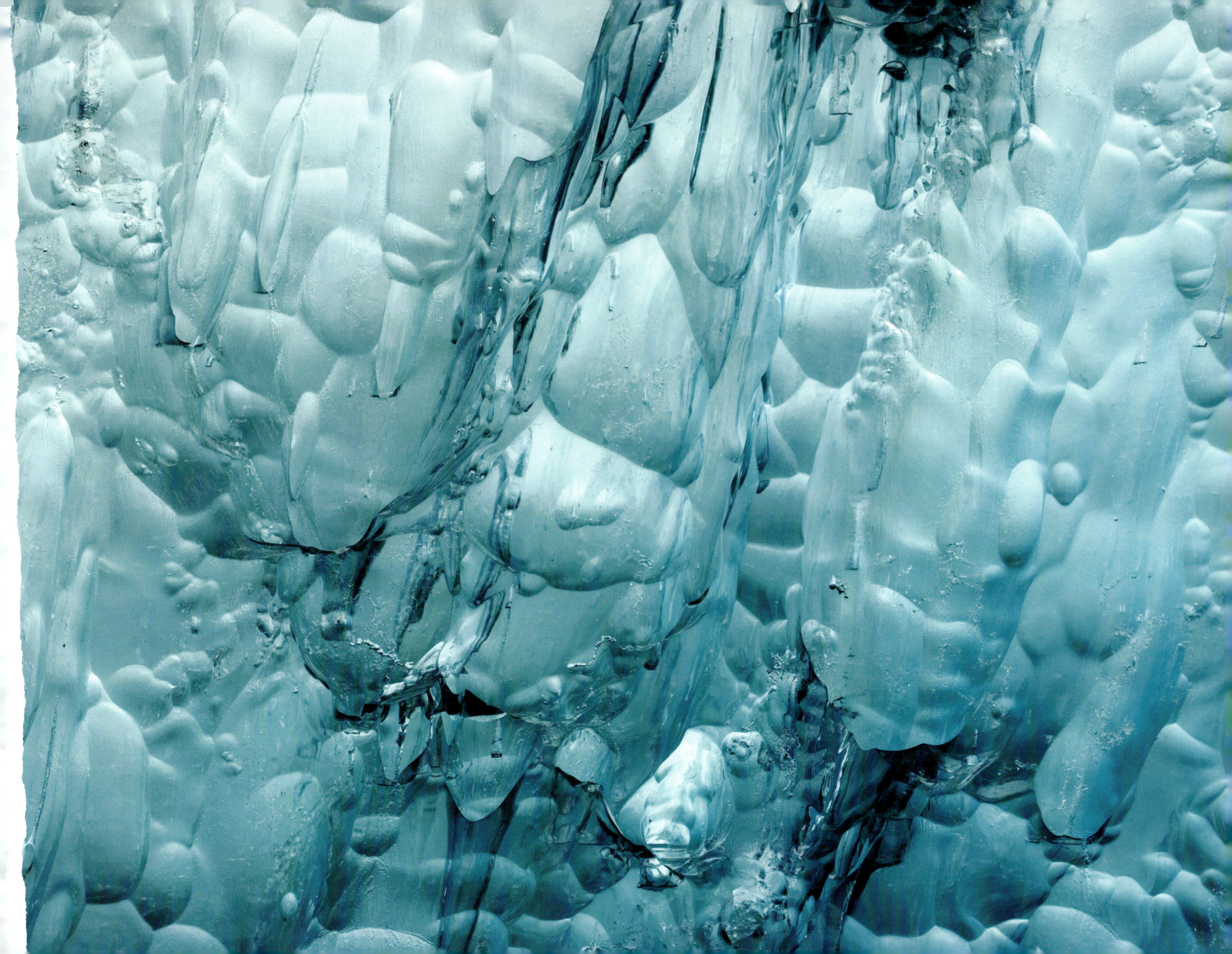

Published by Inhabit Media Inc.
www.inhabitmedia.com

Inhabit Media Inc. (Iqaluit) P.O. Box 11125, Iqaluit, Nunavut, X0A 1H0

Editors: Neil Christopher, Anne Fullerton, and Kelly Ward
Art Director: Danny Christopher
Designer: Sam Tse

We acknowledge the support of the Canada Council for the Arts for our publishing program.

This project was made possible in part by the Government of Canada.

ISBN: 978-1-77227-445-5

Printed in Canada

Library and Archives Canada Cataloguing in Publication
Title: Siku : life on the ice / photographs by Paul Souders.
Names: Souders, Paul, author, photographer.
Identifiers: Canadiana 20220428522 | ISBN 9781772274455 (hardcover)
Subjects: LCSH: Ice—Arctic regions. | LCSH: Ice—Arctic regions—Pictorial works. | LCSH: Ice—Social aspects—Arctic regions. | LCGFT: Photobooks.
Classification: LCC GB2595 .S68 2022 | DDC 551.310911/3—dc23

Pg 48 text: Metayer, Maurice. *Unipkat: Tradition Esquimaude de Coppermine, Territoires-du-Nord-Ouest, Canada*. Québec: Centre d'études Nordiques, Université Laval, 1958.

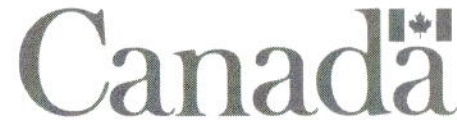

Canada Council for the Arts
Conseil des Arts du Canada

Notes on Inuktut Pronunciation
Inuktut is the word for Inuit languages spoken in Canada, including Inuktitut and Inuinnaqtun. The pronunciation guides in this book are intended to support non-Inuktut speakers in their reading of Inuktut words. These pronunciations are not exact representations of how the words are pronounced by Inuktut speakers.
For more resources on how to pronounce Inuktut words, visit **inhabitmedia.com/inuitnipingit.**

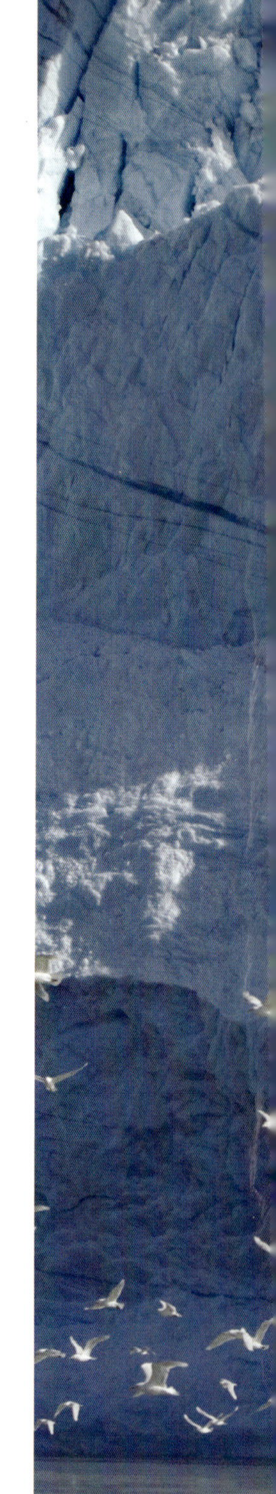

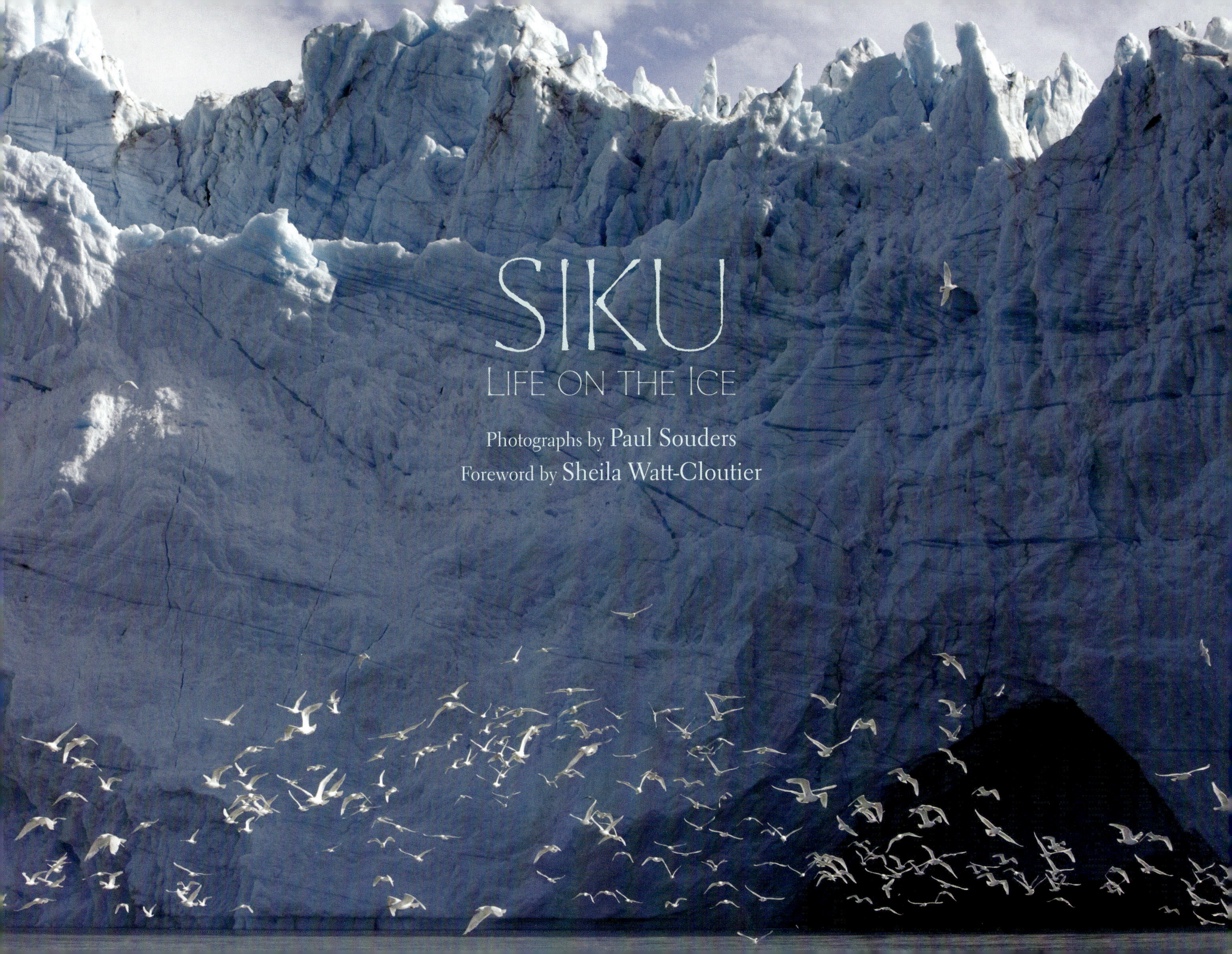

SIKU
LIFE ON THE ICE

Photographs by Paul Souders
Foreword by Sheila Watt-Cloutier

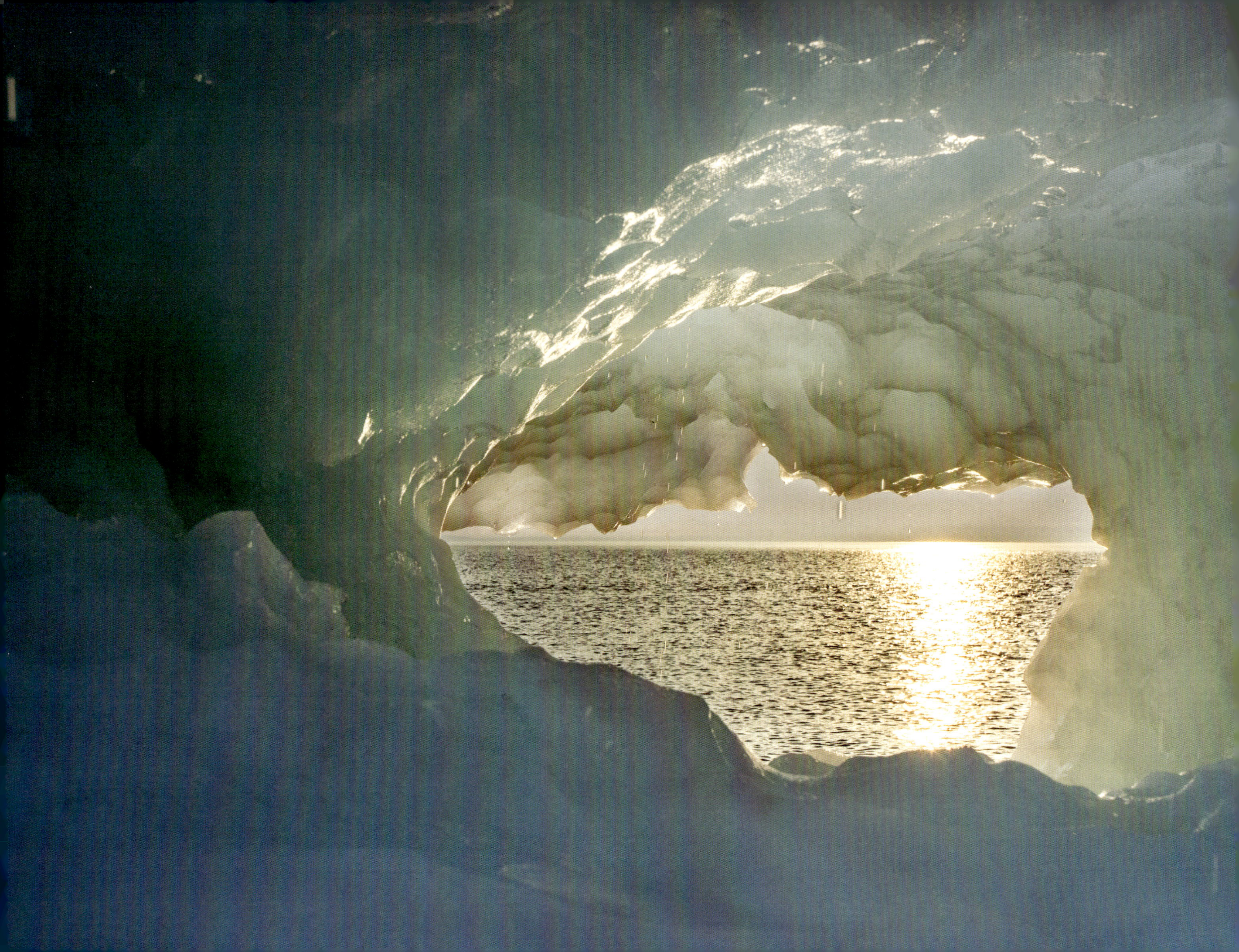

Foreword

This is an incredibly appropriate time to discuss the critical importance of the oceans and ice, particularly in the Arctic.

Many around the world are unaware of the enormous challenges that our Northern communities have faced over the last decades due to colonialism and historical traumas, as the world is aware of the wildlife of the Arctic more than its people. Nor are enough people aware of how important the Arctic ice and glaciers are to the rest of the planet.

My life's work over the last twenty-seven years has been to humanize the issues of the environmental degradation of the Arctic, due to airborne pollutants ending up in our food chain and the devastating impacts of climate change to our way of life, which depends on the cold, the snow, and the ice. The ice is our life force, and we as Inuit are reliant on it for our mobility and transportation. We travel the icy highways to get to our hunting and fishing grounds to bring home our "organic," healthy country food.

Further, the land and ice are what I often call our university. It not only teaches our young people to become proficient providers and natural conservationists, but it teaches them in a natural setting the character and life skills needed to survive and thrive in what is considered one of the harshest environments in the world. When one is waiting for the weather to improve, the ice to form, the animals to surface, one is learning patience. One also learns to be bold under pressure, taking the right survival-based risks. One learns courage, determination, focus, endurance, persistence. In other words, the land builds resiliency and

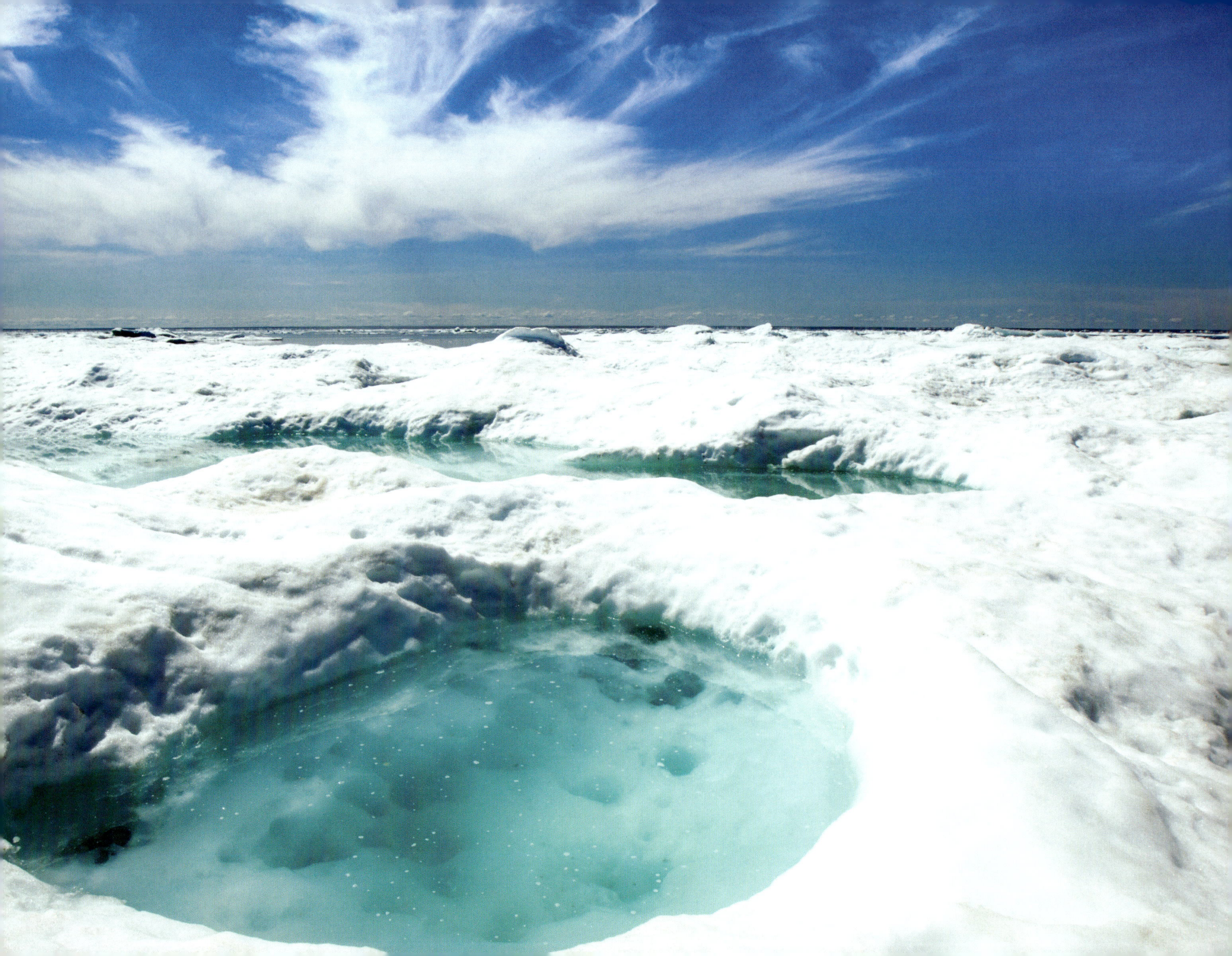

ultimately sound judgment and wisdom. We call this *Silatuniq*, wisdom, which is the hallmark of Inuit teaching and learning. When young people learn to read the conditions of the ice, they are learning about how the world works around them; however, the character and life skills one learns on the ice are also about how *you* work, and how you work is at times even more important in our world, as well as the world outside the Arctic. These skills are very transferable to the modern setting. We have learned that those who have built their character and life skills through traditional means have a built-in capacity to better withstand the modern stressors of our day.

Therefore, I strongly believe the character and the life skills taught by culture on the land and ice are the medicine we need to lift ourselves out of the despair we live in due to historical traumas.

The Arctic has been subjected to the most dramatic environmental effects of globalization. From persistent organic pollutants ending up in our food chain, to our weakened ozone, and, most recently, to the huge changes to our lands and ice from climate change, we have borne the brunt of development far from home, and have been compelled to reach out to the world. The changes—such as permafrost melt, coastal erosion, thinning ice, beach slumping, and unpredictable weather patterns—are just a few of the immense changes happening, creating issues of safety and security for our way of life as Inuit.

If we want the next generation to overcome the health and social challenges we face in our communities, then we must ensure that the ice is intact for them to continue to develop their coping and resiliency skills. The importance of ice for us as Inuit is a human issue. As the ice goes, so too the wisdom along with it.

Silatuniq | SEE-lah-too-neek | wisdom

In addition, on a global scale, the Arctic ice serves as the cooling system—the air conditioner, if you will—for the planet. It is now breaking down, and this breakdown is causing havoc all around the world, showing up as erratic events such as wildfires, droughts, floods, intense hurricanes, and tornadoes. This link is clear. What happens in the Arctic does not stay in the Arctic. If we protect the Arctic ice, we save the planet. This is why we defend our "Right to Be Cold." It is our human right to be Inuit who depend on the cold and ice for our livelihoods, which we defend.

We are now in a place of great opportunity to reconnect to what really matters. That human trauma, such as what we as Inuit have been through and what our planet is going through, is one and the same. Unsustainable ways cause irreparable damage to the atmosphere, and they are forcing the planet to react with violent storms and other erratic events. This is not unlike the Inuk child who has suffered trauma. Without care, a space to heal, and effective coping mechanisms, self-destructive behaviour is inevitable.

We don't want to be just victims of globalization and commercialization. Therefore, it is important to understand these interconnected links to ensure the younger generation can continue to be nourished, healed, and taught by the ice. In turn, they themselves can become the champions for a remarkable way of life, which depends on the ice for the physical, emotional, and spiritual growth and well-being of their people.

As I have often said, the future of Inuit is tied to the future of the rest of the world. We are now part of the global economy, part of global society. And our home is a barometer for what is happening to our entire planet. If we cannot save the frozen Arctic, can we really hope to save the forests, the rivers, the farmlands of other regions?

—Sheila Watt-Cloutier, author of *The Right to Be Cold: One Woman's Story of Protecting Her Culture, the Arctic, and the Whole Planet*

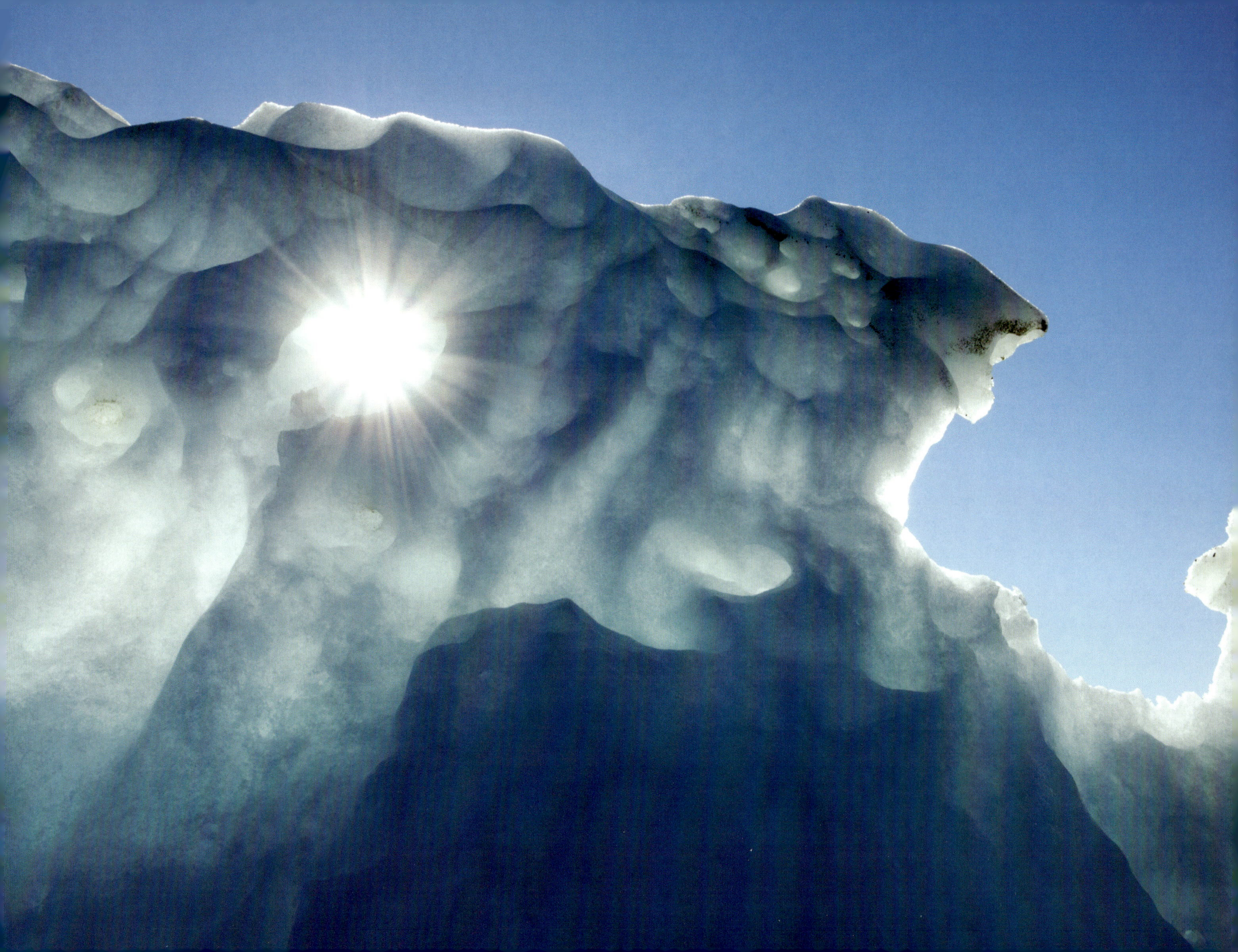

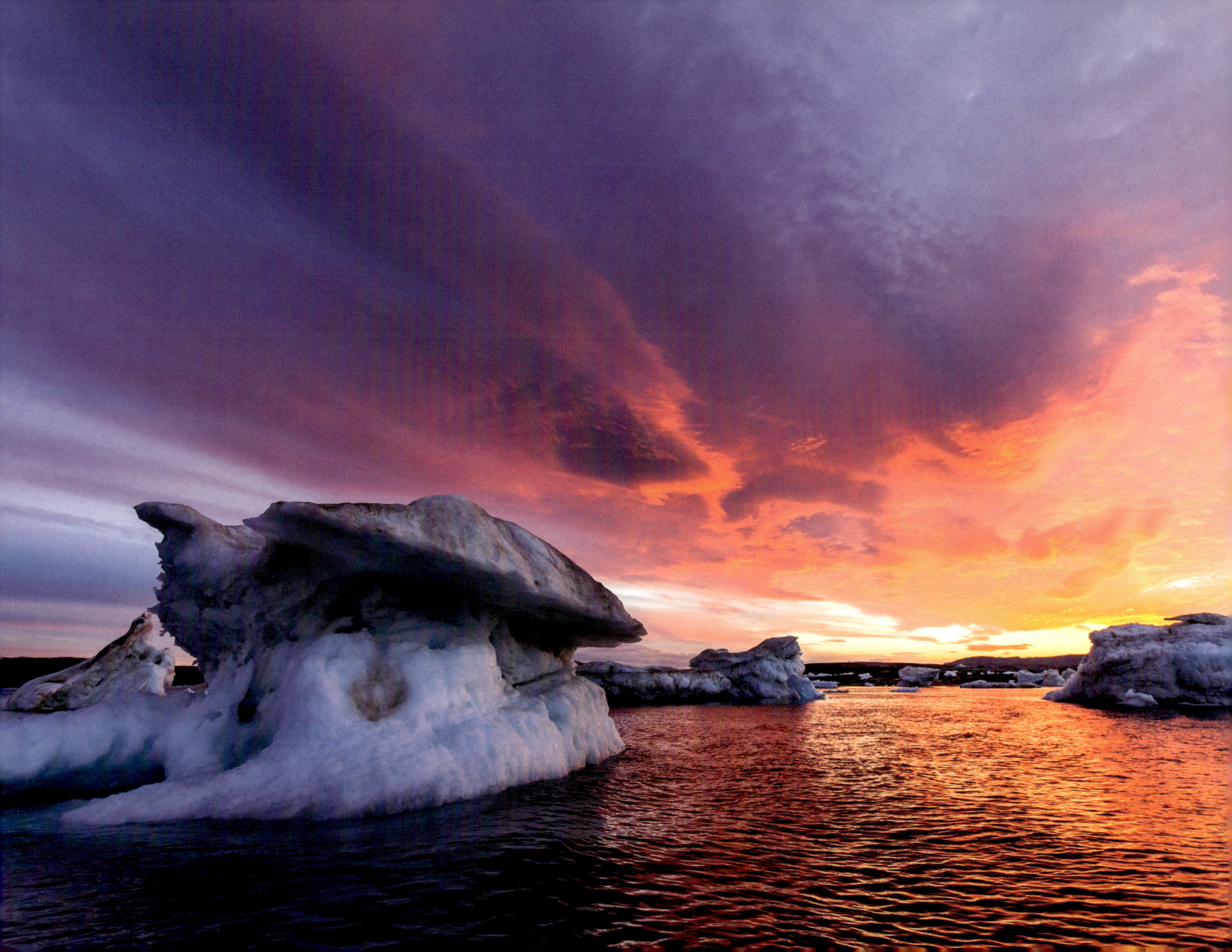

It was early June in 1997, and I was going on a class trip to study marine biology. There were sixteen of us: nine high school students, one teacher, one hunter with a guide, and two Elders with a guide and a child. We left Pond Inlet to meet students from Arctic Bay at the Lancaster Sound floe edge.

On our second day we were near the floe edge and travelling toward Arctic Bay, but as we travelled we noticed a crack opening up and we started drifting. We tried turning back but we couldn't cross the crack in time. My *qamutiik* broke, and myself and four others were stranded from the rest of the group for a while. But eventually we all got together and the Elders said that we needed to find a piece of multi-year ice so that if the ice started to break up, we would be safe on the old ice. An Elder found the multi-year ice, which is more than two years old and is salty. It's usually flatter than icebergs, which can be brittle and tippy. We quickly fixed my qamutiik, with the guide teaching me how to tie the crosspieces properly. An Elder also made a temporary *iglu* with a tarp on top to give shelter to the students. After we fixed the qamutiik we moved to safer ice and set up camp. I made sure my tent was secure, tying all the ropes to snowmobiles and *qamutiit*. We used an HF single sideband radio to notify the community and search and rescue. That Christmas my father had gifted me a GPS. It was the first in the community, and people didn't know what they were. But the GPS came in very handy since people knew exactly where we were. That night got very windy, and we stayed in the tent, horrified that the ice might break up underneath us.

We woke up to very heavy rainfall. It was so weird to see rain early in the season. The rain turned to snow and finally calmed down after two days. Some of us

qamutiik | KAH-moo-teek | sled
iglu | EEG-loo | snow house
qamutiit | KAH-moo-teet | sleds

teenagers ventured out within the one-kilometre-by-half-a-kilometre-wide ice pan. There was an iceberg nearby, so we went on top to scope around. We saw fresh polar bear tracks but didn't see the bear. We could see the land we had come from get farther away, but saw Devon Island getting closer—land we hadn't seen before. We headed back to our camp.

We kept radio contact every hour or so. They were sending a search and rescue team with boats. They also said that a Hercules airplane was on its way to drop off supplies. On the third day the Hercules dropped more food and naphtha fuel for our stove. We still had a bunch of supplies, so we all just grabbed chocolate bars and candies from the military rations. On the fourth day a Sikorsky helicopter arrived to pick us up. We were so amazed that it knew exactly where to go with the GPS coordinates. They had no problem finding us. We had to leave everything behind, but we were told we could take our rifles. We were so happy to be rescued, but sad to leave our equipment behind. There were five snowmobiles in total, most of them brand new, and lots of nice tents and good hunting gear.

We eventually made it home safely.

It is always dangerous near the floe edge, as ice usually breaks up, and strong currents and wind can cause the ice to start drifting. We always need to be aware of where we are and head to a safe area as soon as we find out that the ice is drifting. Always have a SPOT device, GPS, and communication with you on the ice. Check the distance from the ice ridges to the land to see if you are moving. If it's foggy, use the GPS and check coordinates often to make sure you are in the same spot.

—Brian Koonoo, Pond Inlet

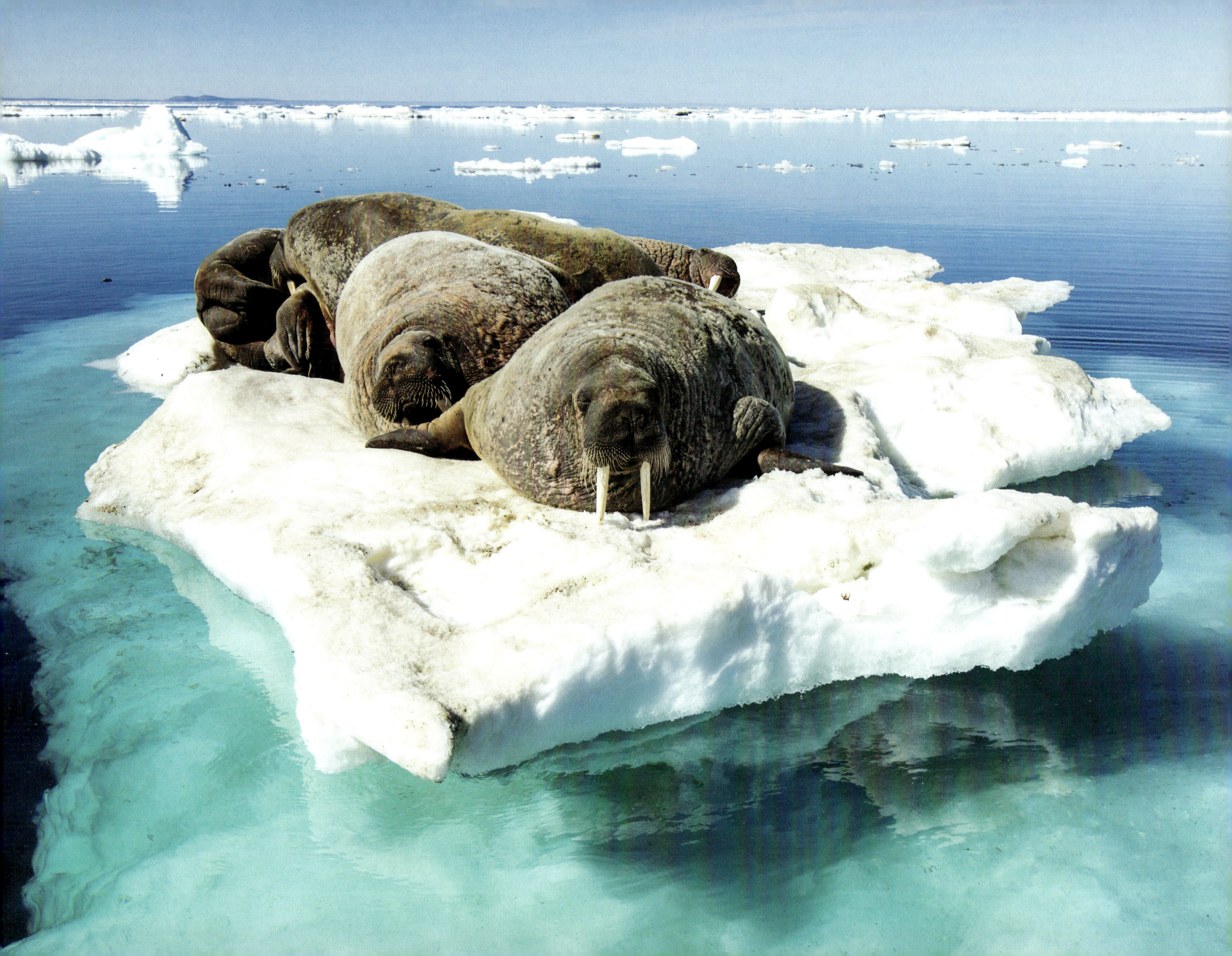

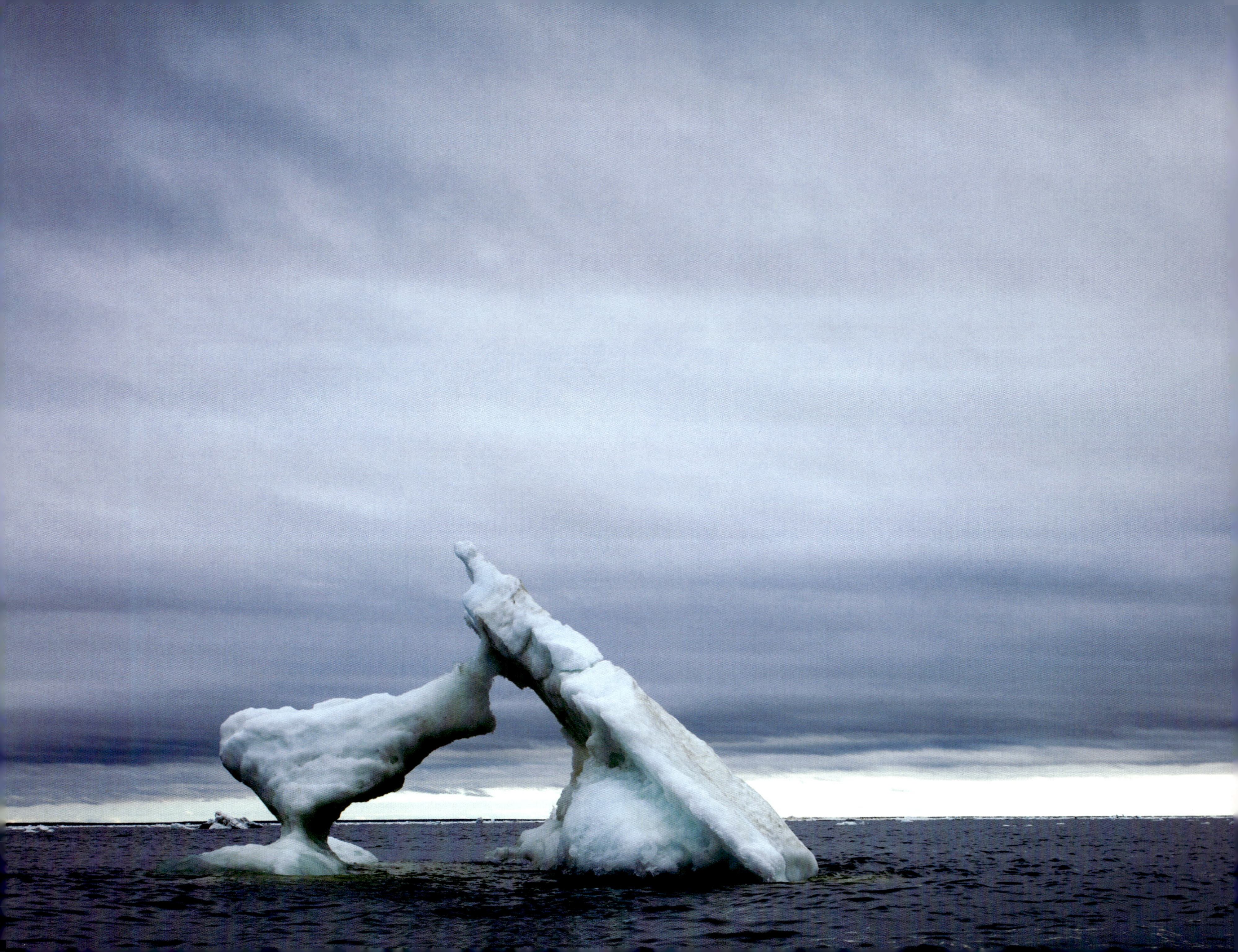

Travelling alone in a twenty-two-foot cabin cruiser named *C-Sick*, I had set out from the town of Gillam, Manitoba, motoring slowly along the banks of the Nelson River, winding up nearly a thousand miles north into Repulse Bay, close to the Arctic Circle and to the village of Naujaat. I had come north to photograph polar bears, and for more than a month, I'd been searching in vain for the dense sea ice that served as the bears' home and hunting platform.

One morning, though, anchored out near Gore Bay just east of Naujaat, I woke up to find pack ice stretched out in front of me, a wall of white. This was the stuff of four hundred years' worth of explorers' nightmares. British Navy ships vastly bigger and stronger than poor little *C-Sick* had been crushed to matchsticks in that same ice. Back home, I had a whole shelf of books that could best be described as disaster porn. Stories of overconfident and underprepared men going off to meet their fates, to starve and freeze and die far from home. I had brought a couple along as bedtime reading. From all that I'd seen, everyone from Henry Hudson, set adrift by mutineers in 1610, to Sir John Franklin, the man who ate his boots on his disastrous Northwest Passage expedition, they'd all come to grief. I didn't need to think twice—I turned tail and ran like a scared dog.

But for days it felt like I was being followed, chased southwards down the coast of Nagjuttuuq island. I finally rowed to shore in my dinghy and climbed a low hill for a better look. To the south, I could see nothing but thick pack to the far horizon. All I could think was, *Well*, that's *not good*, but when I turned around, I noticed the bay behind me was filling with ice, too. In my journal, I wrote:

> *Like pieces of an enormous jigsaw puzzle, icebergs large and small began to assemble themselves into a solid mass, blocking my retreat. I took off at a dead run toward the shore, leapt into the dinghy, and rowed madly*

back toward C-Sick. *On board, I started driving frantically through, around, and over the gathering ice, looking for any way out. Horrible crunches and grinding noises filled the cabin as I scraped through narrowing passages. It sounded like the fibreglass hull was splintering beneath me, and I waited for geysers of freezing water to start gushing in.*

If I got stuck out there, the pack would take me wherever it wanted. I could spend days or even weeks trying to push my way out. A four-thousand-pound boat against millions of tons of ice? I had better odds peeing over the side and trying to melt my way out.

I saw one small patch of sand and gravel on the rocky shore. By this time, I had run *C-Sick* aground more times than I cared to admit, but never on purpose. But I pulled the outboard motors up until they egg-beatered the water and drove the boat right up on the beach. I thought the ice sounded bad, but this was the sound of something dying.

I killed the motors, jumped out, and did the only thing I could think of: broke out the satellite phone and called home. My wife, who has a real job, was working, and I left her one of those "lost-climber-on-Everest-just-calling-to-say-goodbye" voicemails. I remembered to tell her my latitude and longitude, then asked, "Could you do me a favour and look up the boat insurance policy?"

Then, like any good castaway, I went for a look around my little island. There was ice stacked up fifteen or twenty feet high along the shore. I found some polar bear tracks, as big as dinner plates, and followed them up into the hills. When I caught a glimpse of fur, I froze. All I could think was . . . *polar bear*.

And then, a few steps on, I saw her. She was still where she'd laid down to sleep, years before. Now just a scattering of bones, bleached white by sun and wind,

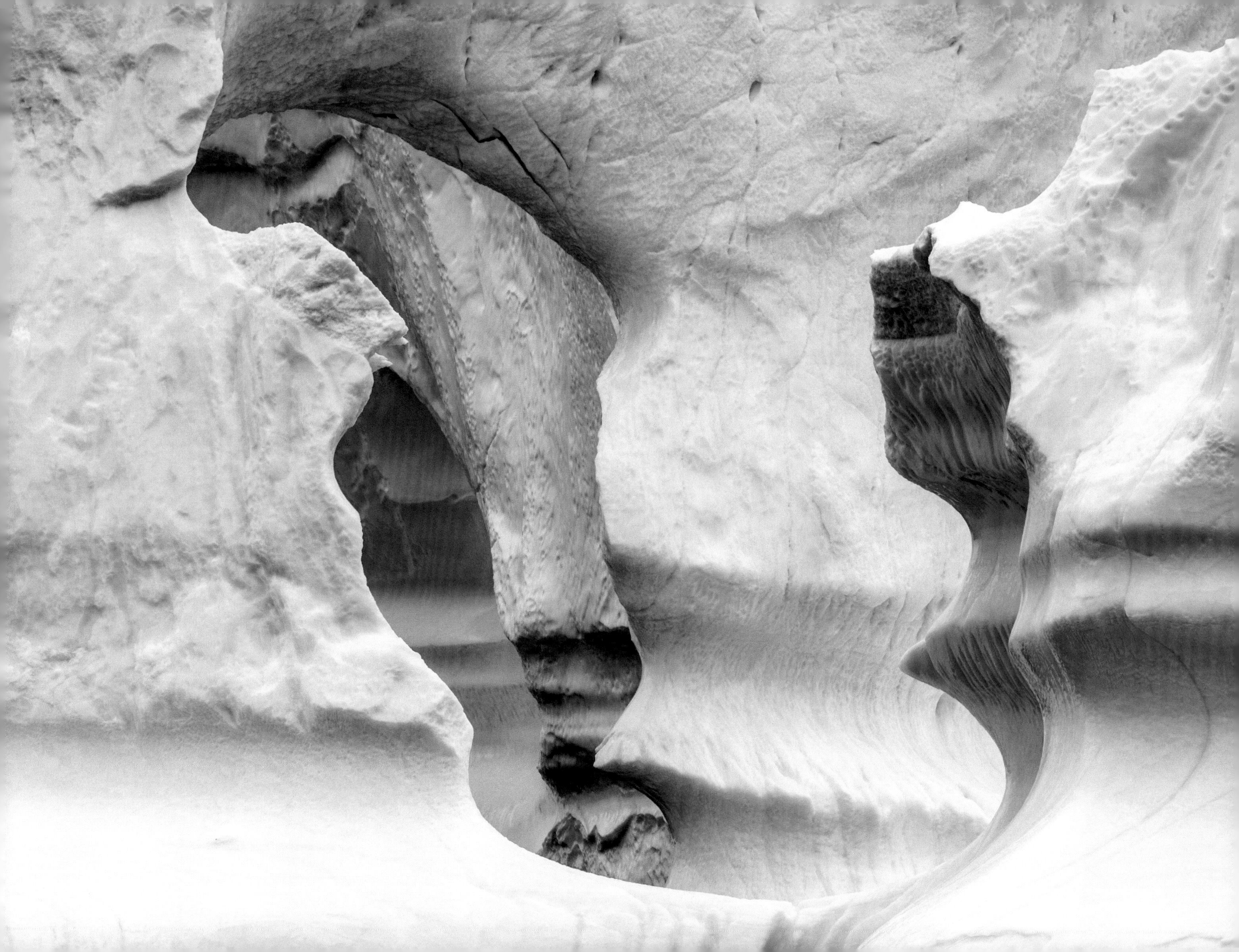

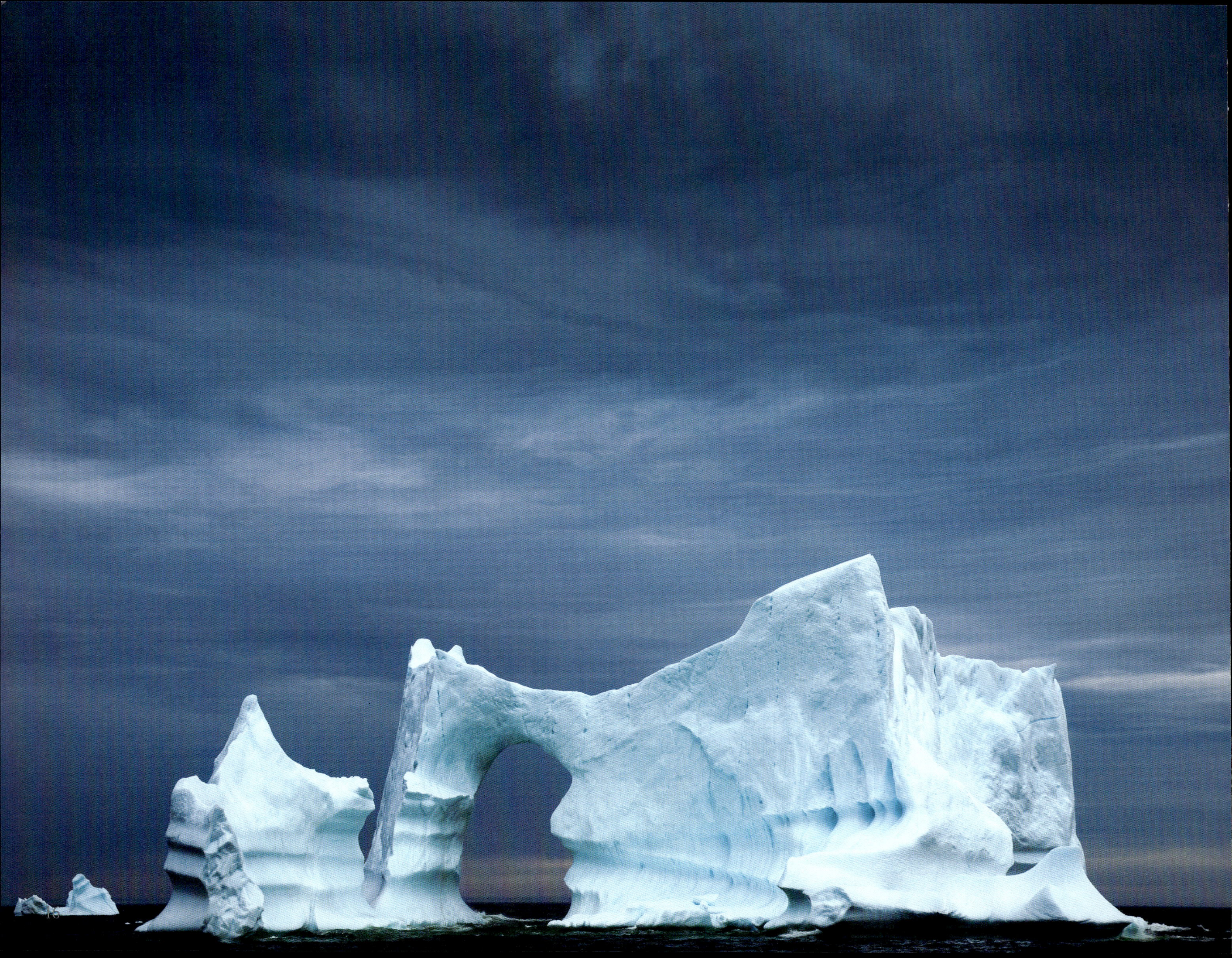

strewn across the tundra. I could feel a cold wind blowing through my bones, too. But I tried to imagine her in her prime: a tough matriarch who could have survived twenty-five winters or more, she might have reared and fed and protected dozens of cubs through her long life. When I finally turned to go, I stopped to glass the ice again.

That was when I saw the other bear. The very much alive one. A couple miles offshore, head held high, sniffing at the air, and walking across the moving ice. Straight toward *C-Sick*.

With the boat high and dry on the beach, I had no place to go, nowhere to hide. If that bear got to the boat ahead of me, or worse, if he caught me out in the open . . . I didn't want to think about it. I set off downhill at a jog, clutching my twelve-gauge shotgun. All this time looking for bears, and now I didn't want anything except for this one to *go away*. I hit the beach at a run, jumped over the gunwale, and grabbed more bear-banger flares. Then I counted out six other shells, the lethal lead slugs I kept locked away, and stacked them in a neat row. Just in case.

A polar bear's nose is the stuff of legend. They can smell a ringed seal at twenty miles through a blizzard. I could see this one taking little sips of air like a wine snob. Spoiled food. Leaky gas cans. Nasty laundry. Then he got to me, Mister Scared and Stinky Pants. Apparently that was too much even for him. He spun on his heels and headed off the way he came.

After sunset, the tide came back in and loosened the ice just enough for me to weave through. When I finally dropped anchor and called home again, the first words I heard were, "Don't *ever* leave a message like that again!"

—Paul Souders, Kivalliq Region

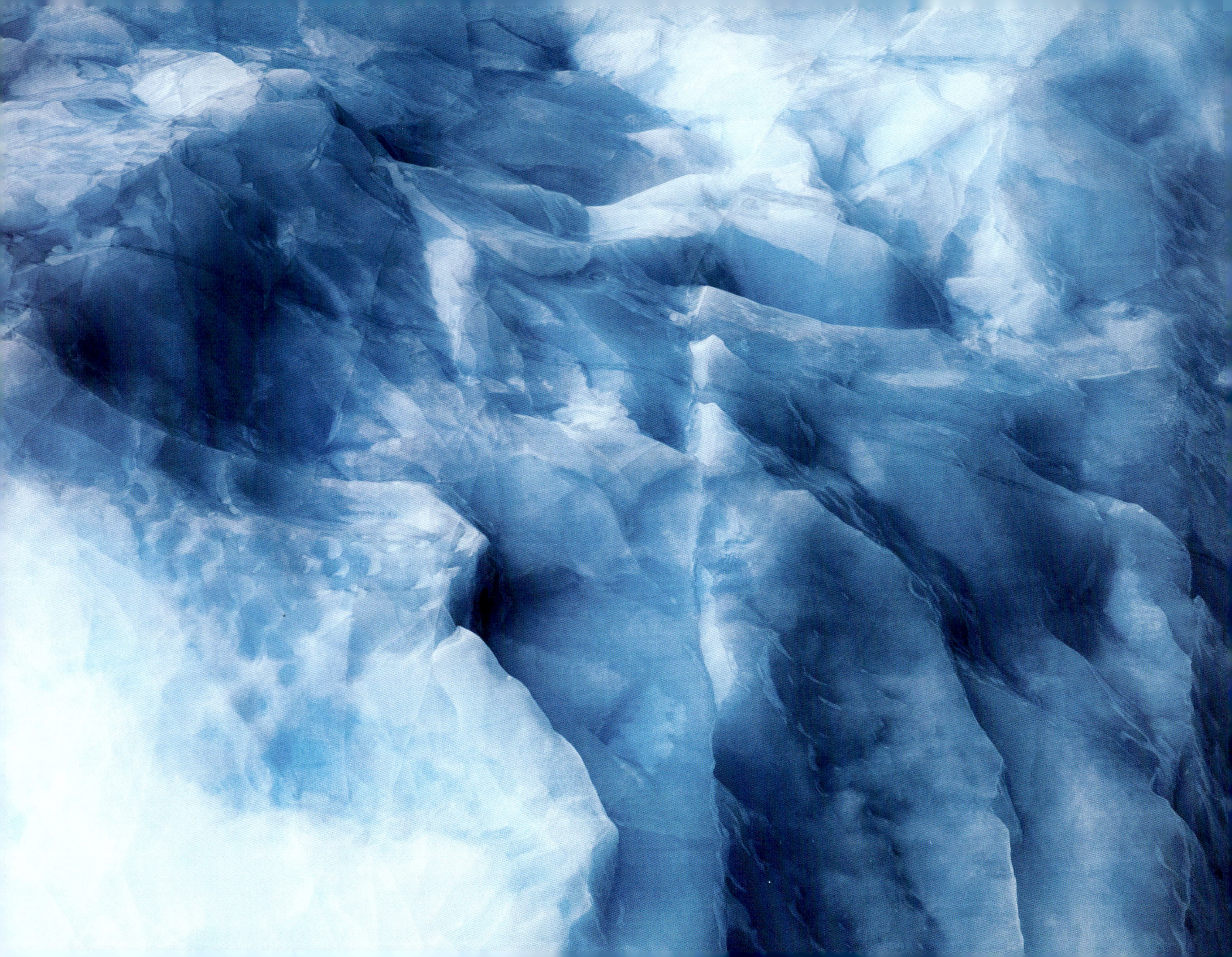

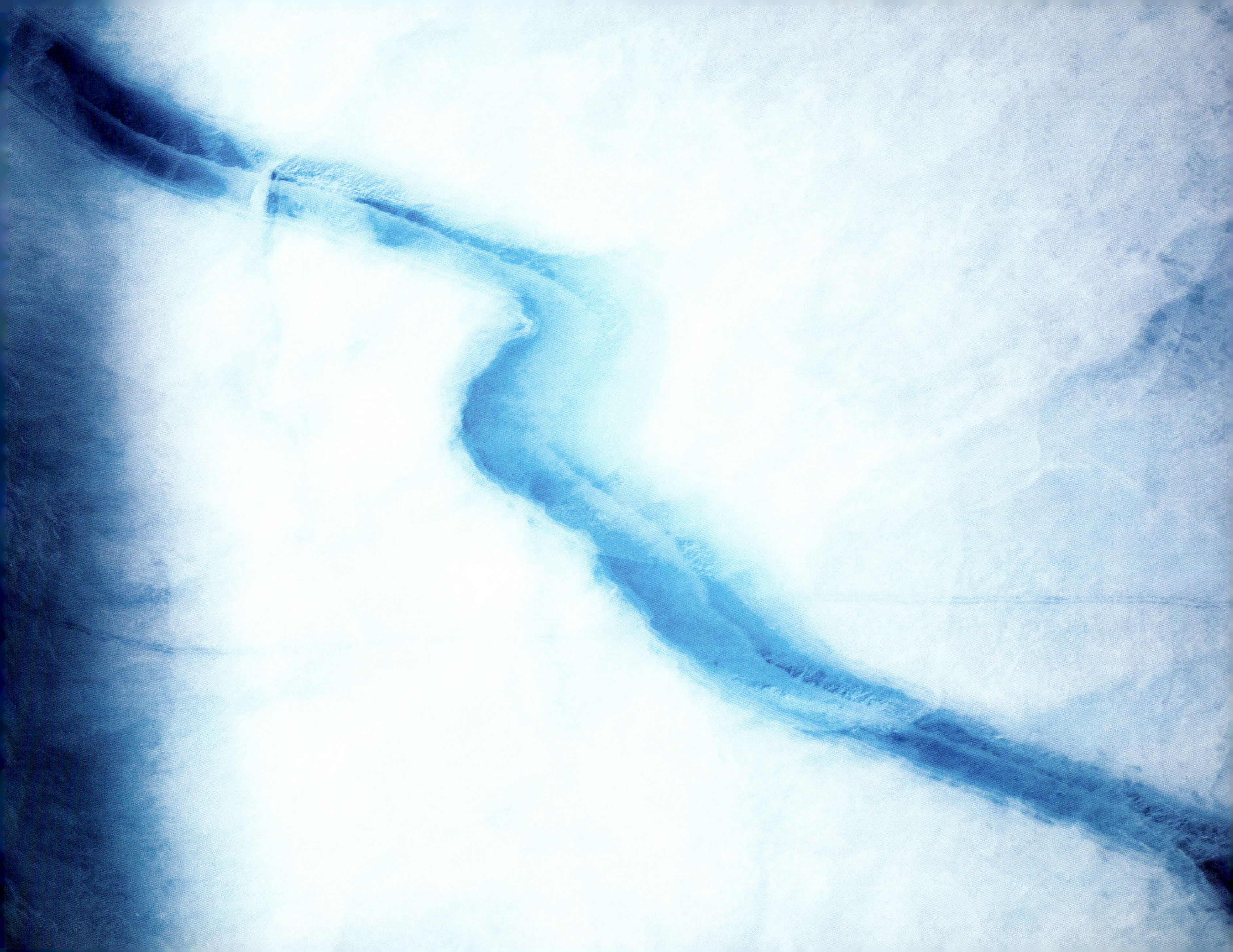

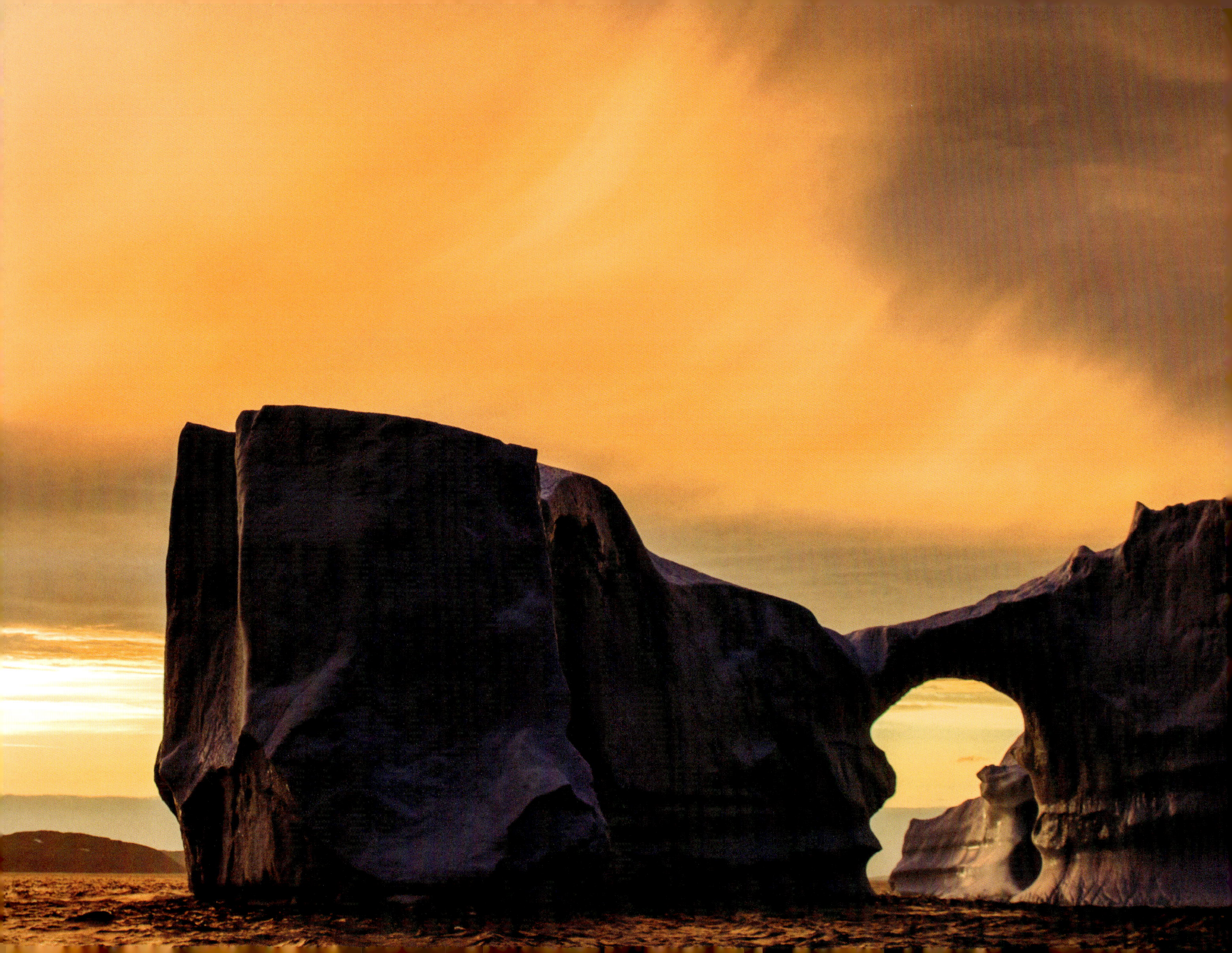

You can walk on sea ice without falling, although it will be moving like flexible plastic. Freshwater ice will break up and you fall in the water. Sea ice doesn't make cracking sounds like the freshwater does. It does not break up, although it moves like a wave when you walk on it, like a large plastic fabric. Freshwater ice is not like sea ice, it breaks up right away while the sea ice doesn't. You must keep walking on sea ice because if you stand still it will break. . . . On one of my seal hunting trips at the floe edge, my partner ran ahead of me on the ice that had just frozen up, and it was moving. He was checking it to see if we could drive the machine through it. The ice was dark, and I was really scared to drive on it. I looked back at my sled and the line of the sled was dragging on the ice because the sled was lower than my machine. He said once I reach the thicker ice it would get rougher. It was cold, so later that day the ice got thicker and safer to travel on.

—Dominic Pingushat, Arviat

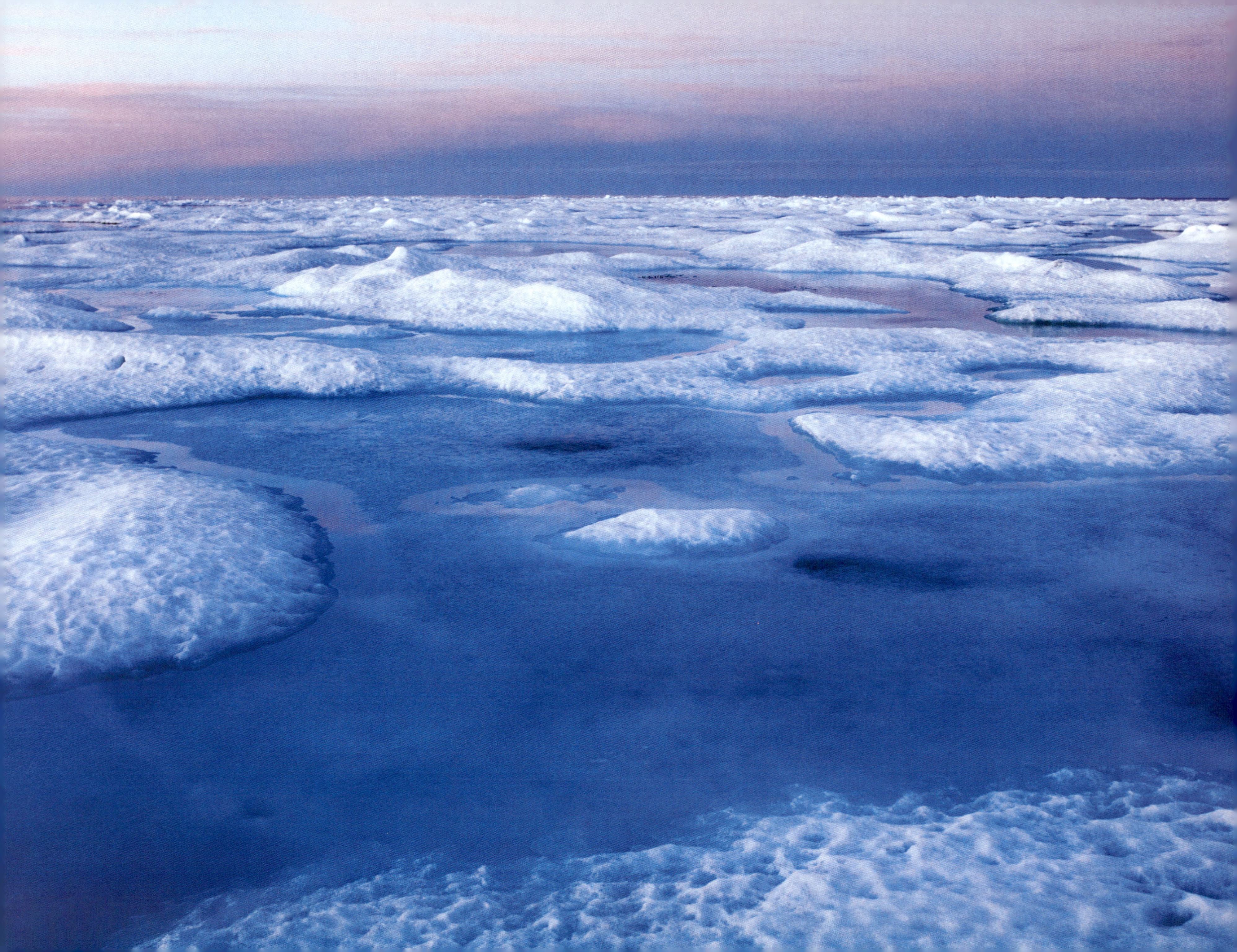

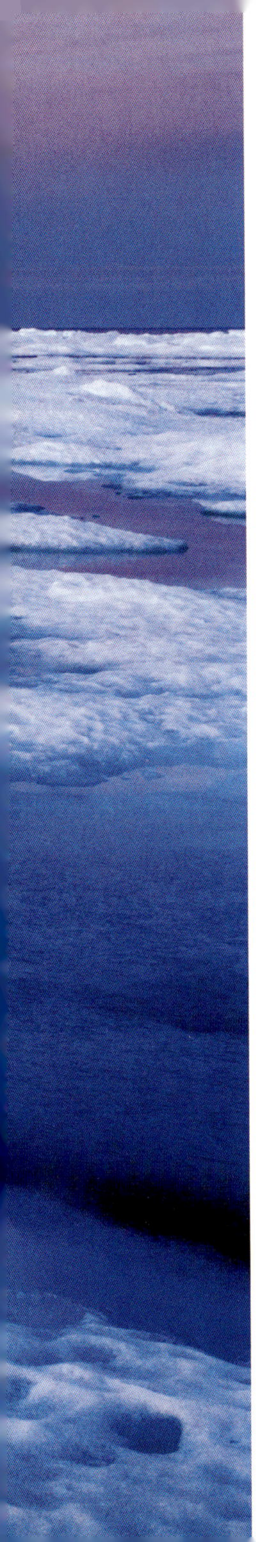

A great deal of hunting is done on the ice. The ice is where you will find animals such as seals, polar bears, fish such as turbot and sharks, clams, mussels, and other underwater species. Also, marine mammals usually hang around the ice most of the year. Birds also are a good source of food and can be found on the ice. Such hunting can be done using a rifle, harpoon, hook, nets, and ropes that will assist you in getting the food.

When hunting at the floe edge, always be on the lookout: watch a piece of ice near the land to see if you are moving. You can also put a harpoon on the ice and use it to see if you are moving. Watch out for polar bears at all times, and check the condition of the ice you are on.

—Brian Koonoo, Pond Inlet

The Canadian Arctic seems like an inaccessible place at times. But what it lacks in roadways, it makes up for in freedom to move across the land uninhibited by fences and other artificial barriers. It was this freedom and desire to move across vast landscapes and learn about new ways of living that attracted me to the Canadian Arctic, and Baffin Island in particular. Most of the year the land is snow-covered, and waterways become frozen highways for both animals and people.

My preferred method of travel is by dog team. It slows the time down. It lets you absorb your surroundings at a much more comfortable pace than zooming by on a snowmobile or other type of motorized transportation. For me, it was also the best way to learn about ice safety and the new environment I found myself in when I moved to Iqaluit six years ago.

Most of my learning has been through trial and error, and as I started raising a team of Inuit sled dogs, we embarked on a joint journey of learning how to navigate the Arctic landscape, including travelling on the ice. I quickly learned that engaging all of my senses, past experiences, and other people's advice, especially related to traditional knowledge, would be key to my enjoyment and safety.

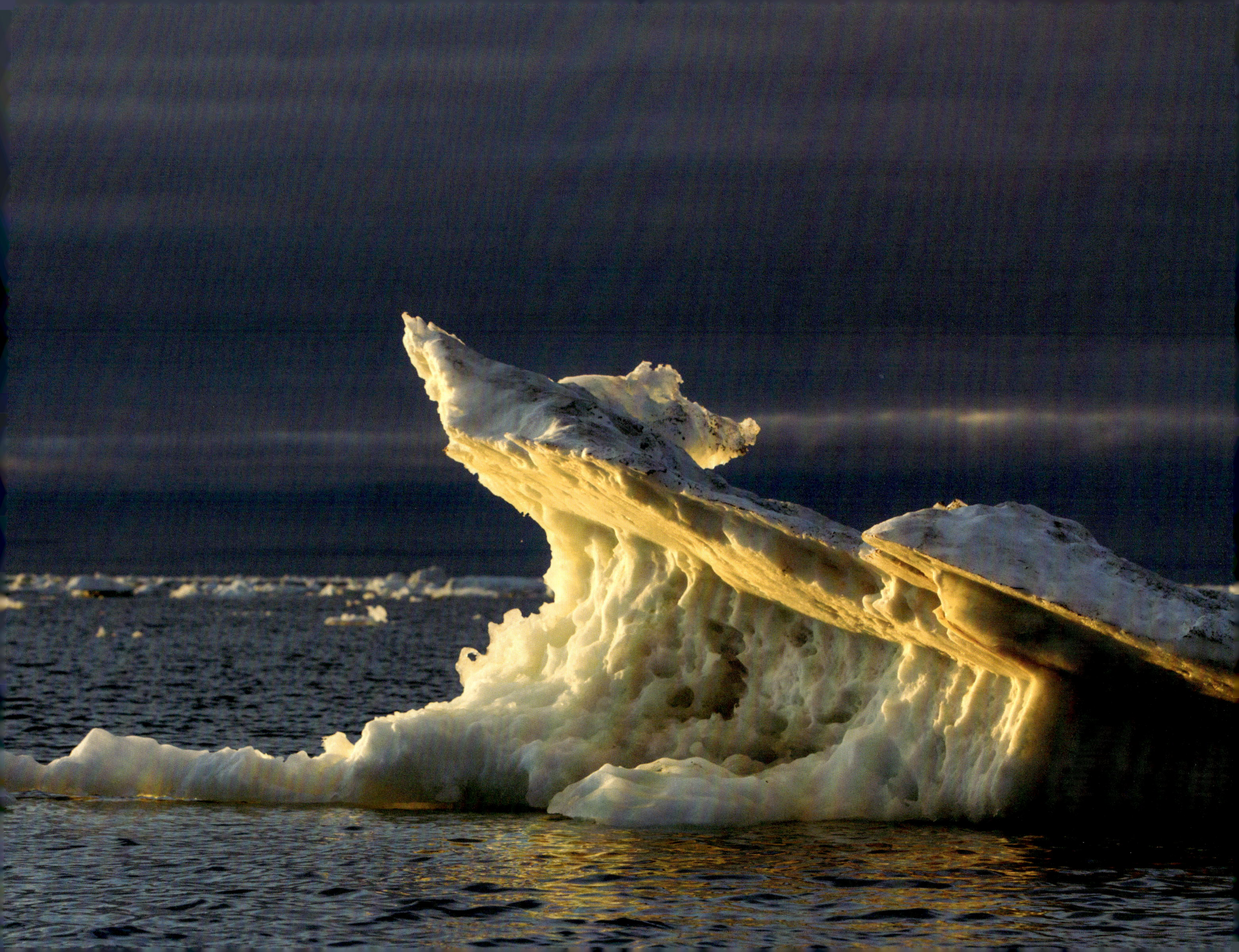

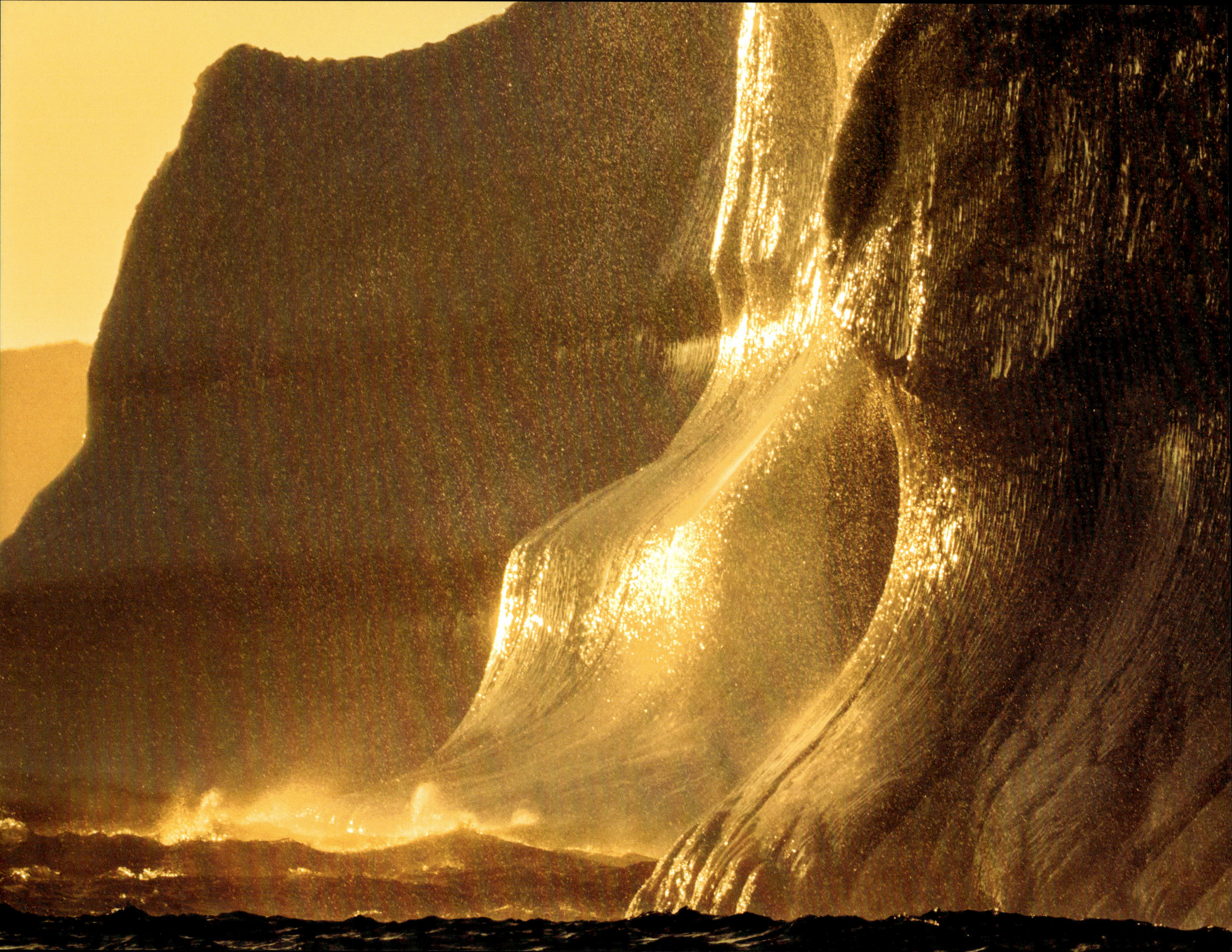

During my first winter in Iqaluit I'd been mostly running with my dogs on the land, but as part of our regular route to get out of town we had to cross Sylvia Grinnell River. By January of that winter, I had already crossed the river a dozen times and didn't think much of it by the time temperatures dipped well below -30°C. That was about to change on one cold January day, and I was about to learn my first ice lesson.

My car thermometer was reading -35°C as I was driving to my dog yard to get ready for the run. We went through the same motions we did every other day: I harnessed my dogs, got the sled ready, and off we went. I didn't pay too much attention or notice any differences in the appearance of the river that day, and as we approached the river the dogs did what they usually do and sped up down the riverbank. In no time we found ourselves in the middle of the river, except this time the dogs froze and started walking like they were on eggshells. One by one they started punching through the river ice, quickly finding themselves in the water, trying to turn around. I tried getting off the sled to help pull them up onto the ice, but I found myself going through up to my chest, feeling the river current at my feet trying to knock me off balance. I was surprisingly not cold at that moment, but I was well aware that that was a temporary comfort and that I would soon have to deal with my cold reality. I crawled back up onto the ice, carefully distributing my weight so as not to punch through it again. As I crawled back to shore, I could only watch my dogs congregate around the *qamutiik*, which was still on top of the ice, but now cemented in the surface slush that had been created through a mixture of river water, snow and ice, and very cold temperatures. At that point I made a decision to run back to town before I got too cold and seek help to retrieve my dogs and the sled. A few friends from the dogsledding community and the local territorial Parks office were quick to respond to my messages, and we were able to get the dogs out of the river and retrieve the rest of my gear.

As I was sitting in the warmth of our home with my friends at the end of the day, we were laughing about the whole experience, and I also learned that I was not the first person that this has happened to. I learned that extremely cold temperatures do not guarantee safe ice conditions. The moon cycle also influences water, like tides and river levels. In this case, the river was overflowing, which created a flow of water on top of the ice that was already formed over the winter. Due to cold temperatures, the overflow itself started to freeze, creating an illusion of safe ice to travel on, when indeed it was just a thin film of ice. Had I been paying attention, there were signs that indicated these changing conditions, like steam coming off the ice, which is due to the fact that the water is warmer than the ambient temperature.

—Jovan Simic, Iqaluit

qamutiik | KAH-moo-teek | sled

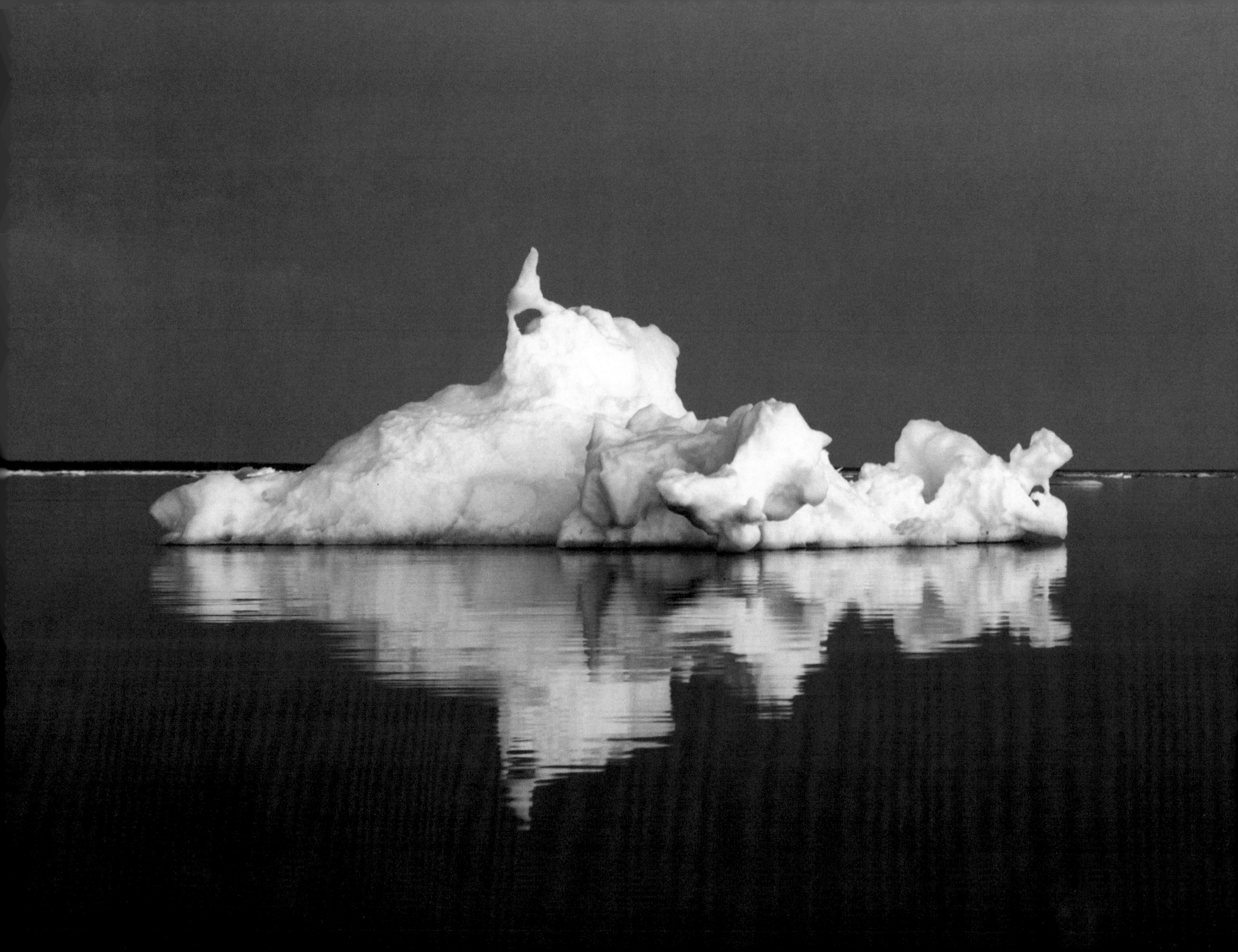

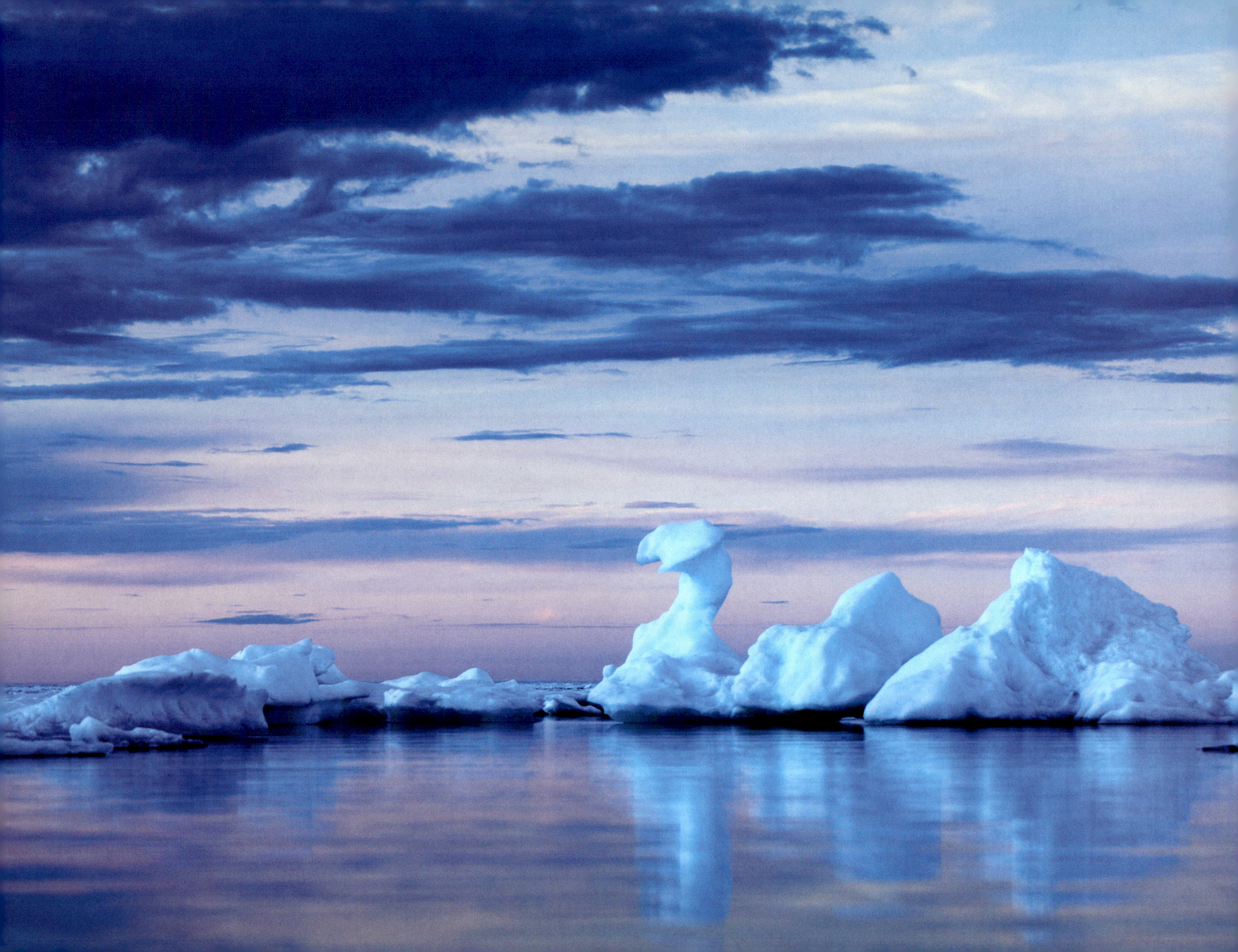

If you fall into the water through the ice, you must fall flat, right away, and roll over to the thicker ice. You may be wet, but at least you will be out of the water. As soon as you know you are falling through the ice, fall forward and roll over. It is harder to get out of the water on fresh water because the ice keeps breaking off when you try to get on it. The water spreads quickly around the hole and it gets slippery and hard to find something to hang on to. There is nothing to grab to pull out of the water. . . . I have fallen through the ice in the river and the lake, and I barely survived. You have to think quickly and act fast. When I fell through the ice, it was dark and my machine went under the water and my sled was in the water. I pushed my sled to the thicker ice and pulled myself onto the sled and ran on my sled to the thicker ice and rolled over. I then took my stove and dried up my clothes throughout the night. I waited until daylight to pull my machine out and drained the water out from the engine and it started.

—Dominic Pingushat, Arviat

Seal holes are easier to notice when the ice is new and without snow on top of it. It is more fun seal hunting when the new ice is formed in the fall, and when it is safe to travel on the ice. The holes are visible even when they are far away from you, because they are raised up from being splashed and then freezing. When a seal is forming a hole on new ice, they splash the water upwards and the water freezes, forming a piece of cone-shaped ice. Of course, not every bump on top of ice is a seal hole. But that cone-shaped ice can be recognized by a hunter as a seal hole. The hunter will approach the cone-shaped ice on top of newly formed flat ice, and sure enough, it will be a seal hole! When the snow falls and covers the ice it also covers the seal hole, and then it is harder to find the hole, and harder to harvest seal—unless you know where to go.

My father, myself, and my brother once went out seal hunting on new ice, before the snow had fallen to the ground and covered the ice. We knew the area and were able to see many seal holes, and were able to harvest some seals that day. A week went by, and during that week it was snowing. My father wanted to go out again to the same area we found the many seal holes. But the thick snow had covered the sea ice and seal holes. We were not able to find many seal holes because we could not see the cone shape anymore.

—Solomon Awa, Iqaluit

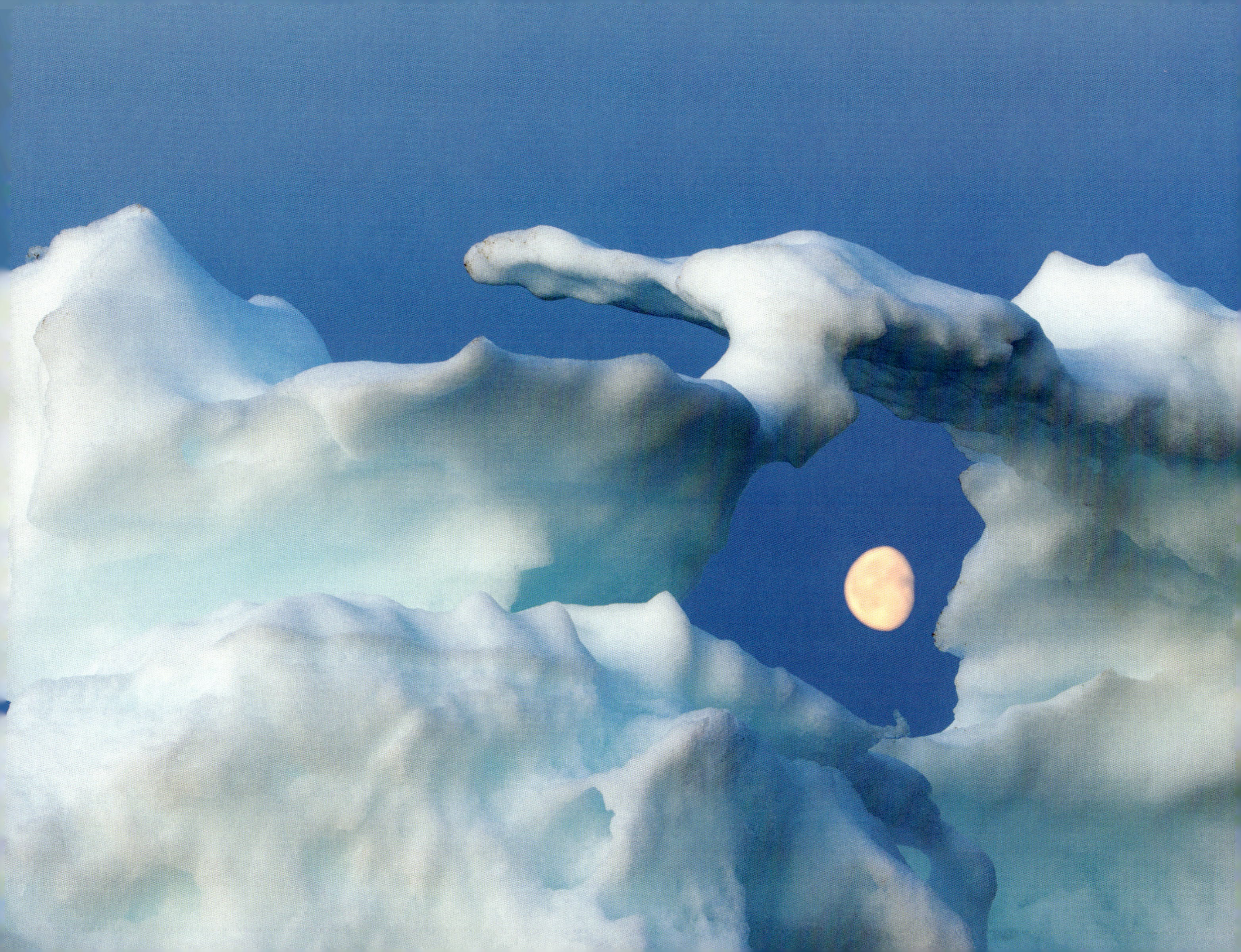

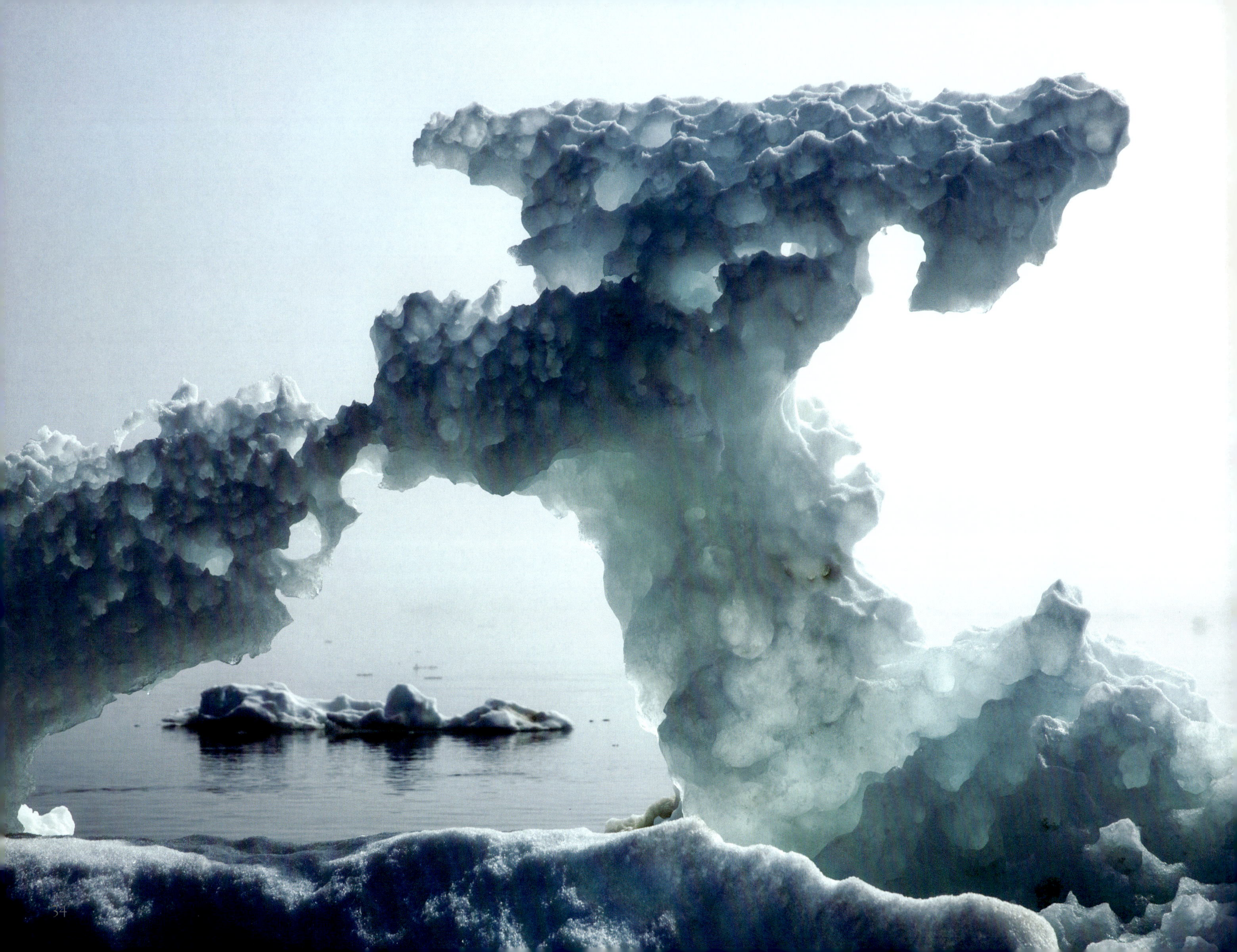

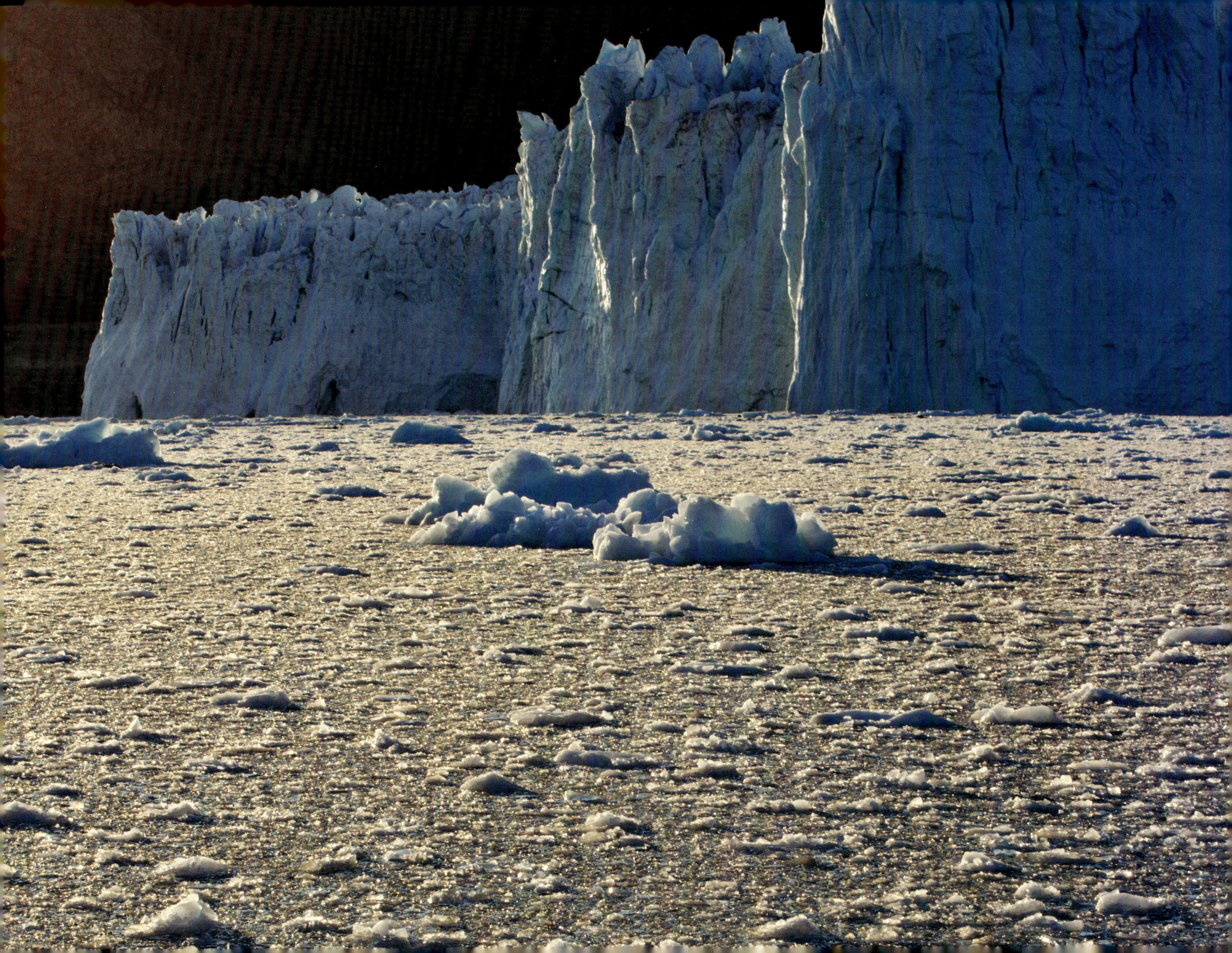

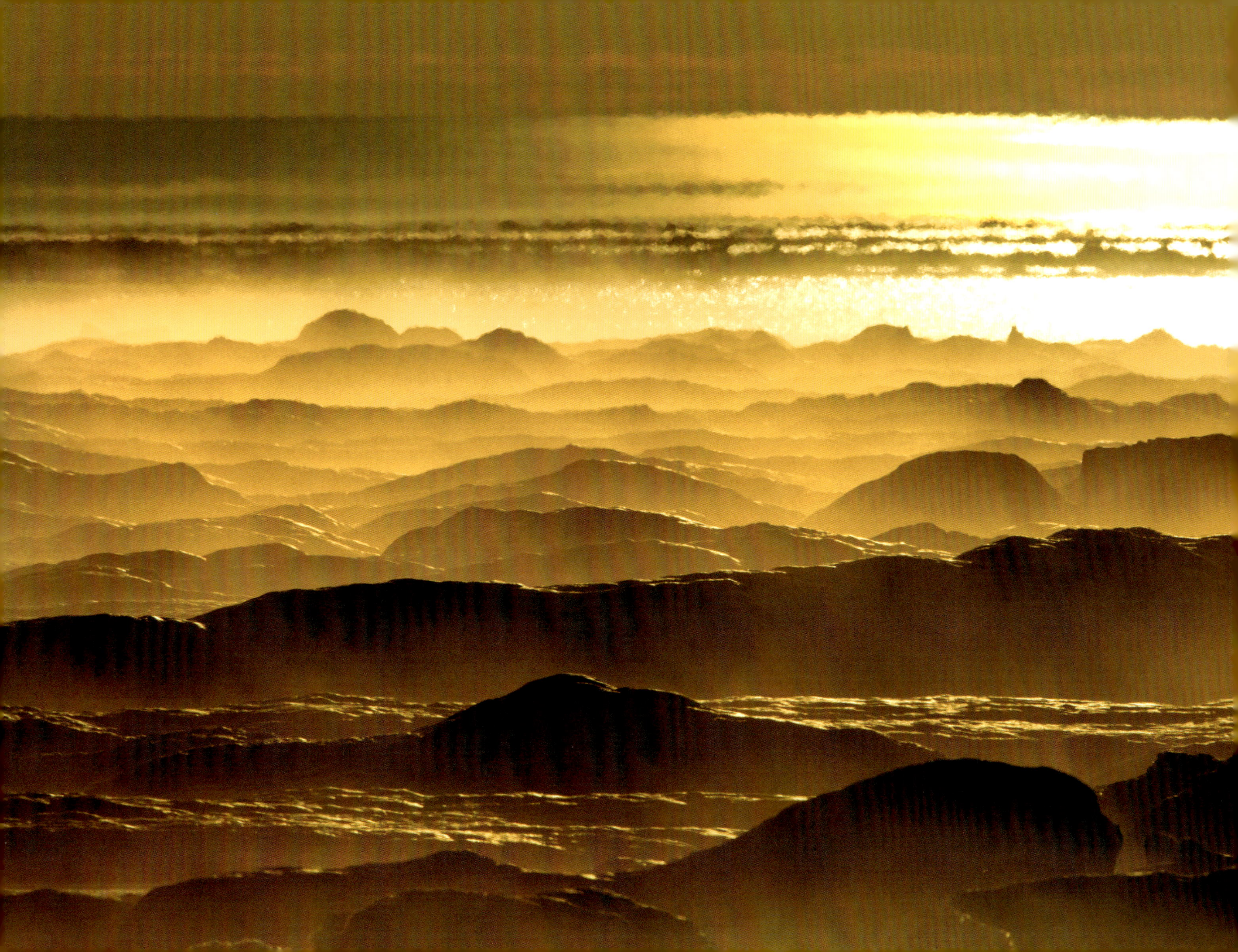

The ability to safely travel on sea ice depends on the time of the year. Sea ice forms according to temperature, snow cover, and current. During winter and fall the ice is safer since it gets thicker each day, and open water can freeze overnight and be safe to cross. But in the spring, when the ice is melting, it's more dangerous.

During freeze-up the temperature is warmer, and it takes a while for the ice to thicken. Salty sea ice is a lot more flexible than freshwater, and it can bend and not crack or break instantly. Once it forms, snow covers and insulates the ice. It also pushes the ice down and water can seep through and make the snow wetter. Sea ice freezes at about -1.8°C, compared to fresh water, which freezes at 0°C. There can even be enough salt in the water to melt snow and ice.

At any time of the year, checking the thickness of the ice can be quickly done with a harpoon. One strike with a harpoon means the ice can hold the weight of a human, but is not recommended for safe travel. Three strikes with a harpoon means the ice is safe to walk on. Five strikes can hold the weight of a snowmobile travelling on the ice. Ten strikes can probably hold a vehicle, and it's safe to travel anywhere depending on the roughness and snow cover.

Travelling on sea ice poses dangers of being drifted out to sea, especially near the floe edge. The ice along the floe edge breaks up frequently, and large pans sometimes start drifting out to sea. The ice along the floe edge sometimes collides and breaks up to form pressure ridges. This happens more frequently during peak tides and windy days. There are certain things to keep in mind when travelling at the floe edge. Sometimes the ice breaks at the floe edge and you can get stranded. In the event of being caught on a drifting ice pan, travel quickly in the direction of where the wind and current are going. There might be a short opportunity to cross safely to landfast ice. If it's not possible to cross safely, look for stable ice or multi-year ice and make camp. This thicker multi-year ice will not break up easily when drifting to the sea, and the ice may eventually drift to landfast ice, where you can cross safely and make it to land.

All in all, the more experienced you are, the more you will be able to travel safely and faster on the ice, and distinguish the dangers that are all around you.

—Brian Koonoo, Pond Inlet

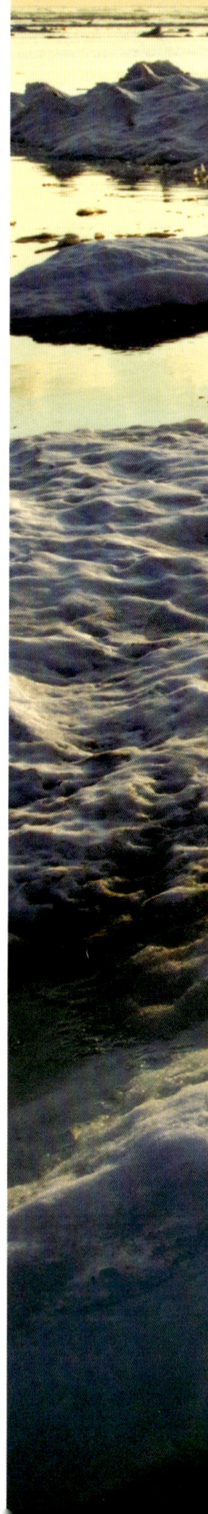

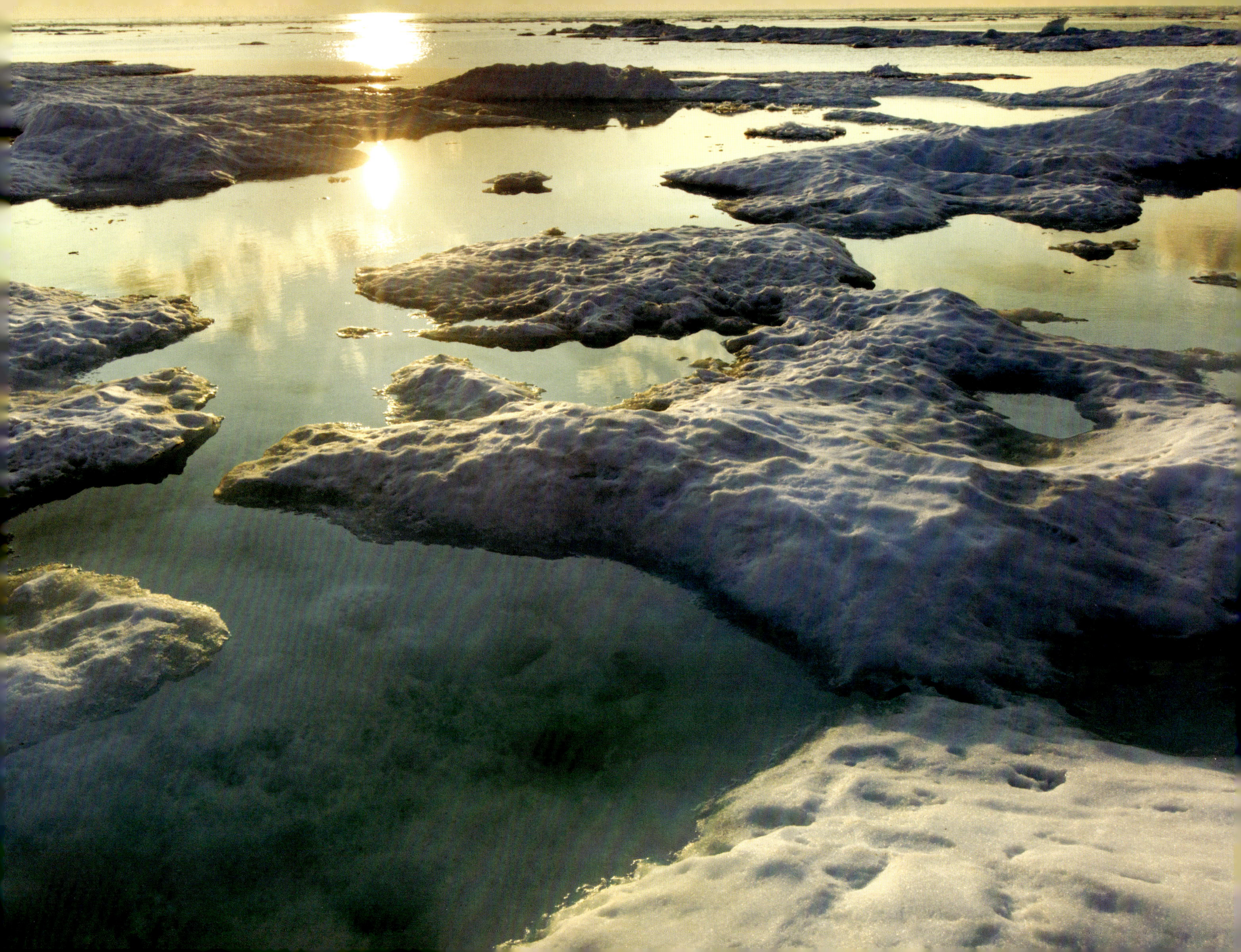

When I was young, there was an abundance of mammals, and we used to catch many Arctic char and seals down at the bay here, right in front of the houses. There would be hundreds of seals basking in the sun. But now, there are none. Some people, committee members, think it is due to climate change, that the weather has become warmer. Animals that were not in this area have increased in number. There are whales, and we have so many polar bears now. My father once told me when they were camping during the summer at Ivvik, farther south from here, that there were no animals around. A polar bear cub swam ashore, and that was the only animal they saw.

—Tommy Ublureak, Arviat

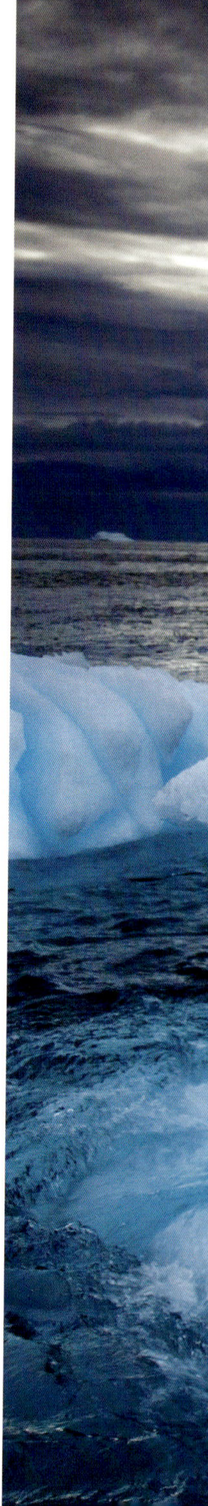

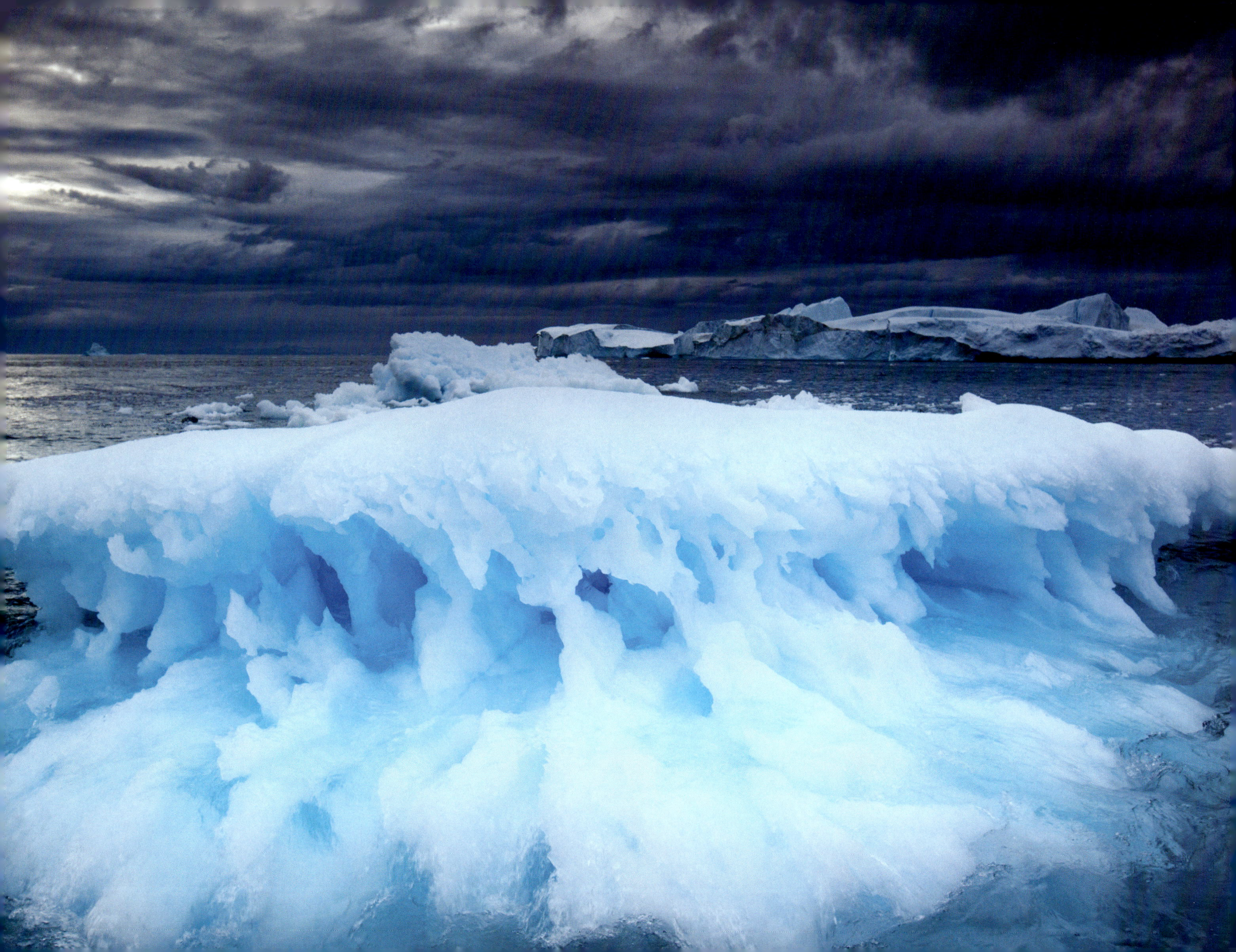

Food is not the only thing that is plentiful on the sea ice; water is also present and easy to get by melting the ice. Once melted, the ice can provide drinking water. Melted ice yields more drinking water than melted snow, and the taste is usually preferred. Freshwater ice can be taken from icebergs, and the tops of multi-year ice where the salt water has drained down to the lower part of the ice. Either way, water will be available and can be used for drinking or for cleaning up a mess after butchering on the ice.

—Brian Koonoo, Pond Inlet

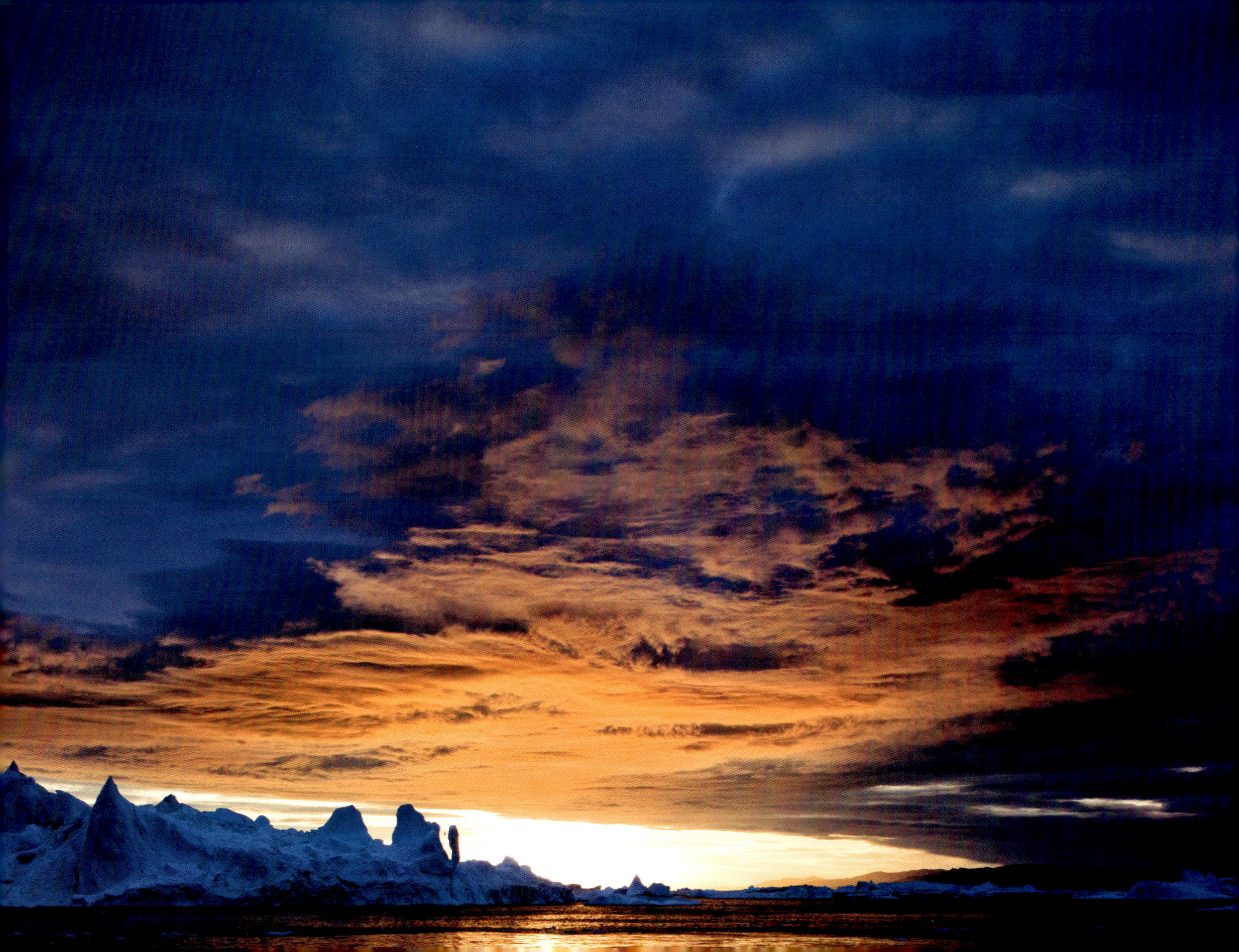

Many of us travel on the sea ice for many reasons during the winter season—even into the spring. We start travelling when the ice is formed and safe, and use it until it starts to melt and is no longer safe. But before it melts, we are able to go to the floe edge during the spring to wait for narwhals.

When we wait for narwhals, we camp at the floe edge, setting up tents on the sea ice. We take turns staying up to make sure that there is no polar bear approaching during our sleep, and of course we are waiting for the narwhals to show up. There is another reason for people to stay up all night and watch the surroundings—to make sure that the ice is not breaking up.

Once, a few people and I were camping at a spot on the floe edge. It was my turn to stay up during the night, but I was not alone, as one of my friends stayed up with me while the others slept. During that time the seawater was very calm and flat; birds were flying by, minute by minute, and some were swimming by our camp. We set up a *qamutiik* to lean against a small boat so it became an A-shaped ladder. I sat on top of it with binoculars, watching the calm, flat-as-glass open seawater, looking for whales approaching.

I spotted one, and I said to my buddy, "I see a whale far way!" He was inside the tent having tea and bannock. He replied, "Don't worry, it might get close soon." He did not even come out to check the whale. I said to him again, "It is coming this

qamutiik | KAH-moo-teek | sled

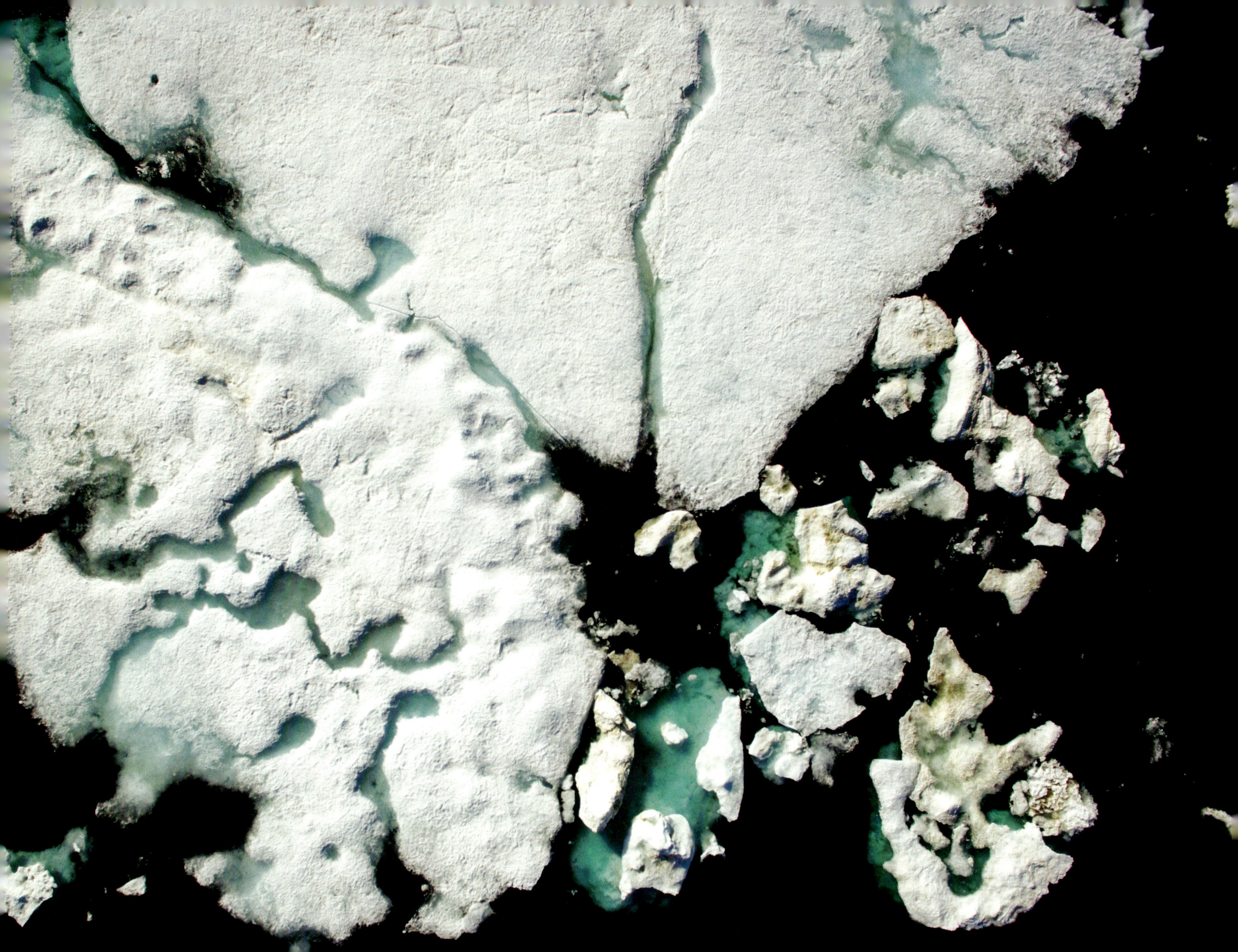

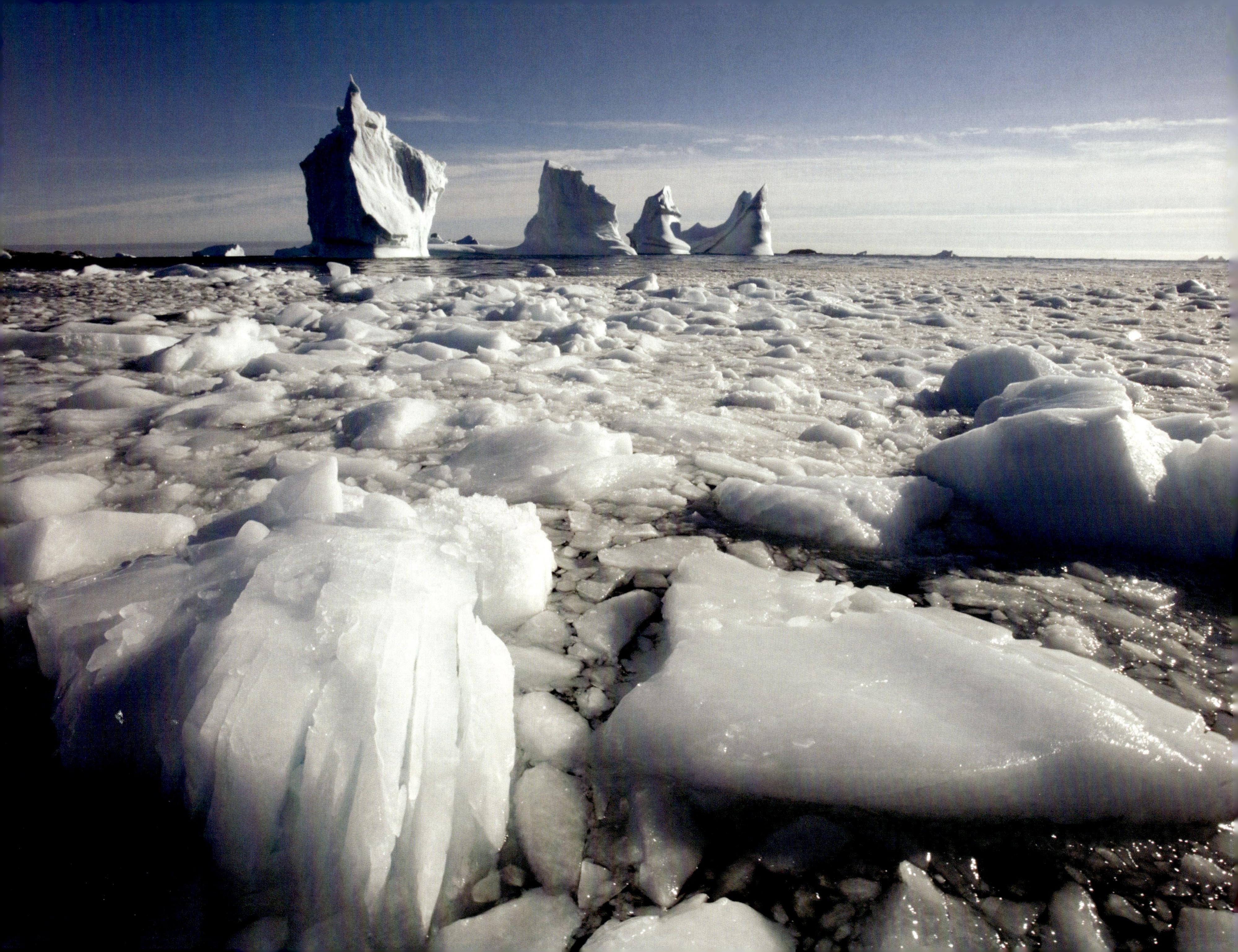

way." "Maybe you can shoot it," he said, still in the tent. I saw another and I said, "I'm going to go check this one, it might get close."

I climbed down the makeshift ladder and took my rifle and started walking toward the floe edge and toward the right side of our camp, thinking that the whale might come up to breathe in that area. As I was still walking it surfaced close by, so I ran to get closer. It popped up once, and popped up again—I measured the distance between where it popped up and took aim at where I thought it would pop up next. It did. I took a shot and hit it, and it started to float. I shouted, "Got a whale, got a whale!" and started running toward the camp, where there was a small boat to retrieve the floating whale. I grabbed the boat and rowed to the whale, retrieved it, and brought it back to the camp.

—Solomon Awa, Iqaluit

It probably happened before I was born. A group of seal hunters were out on the ice when it broke loose from the shore and a thick vapour filled the sky. Ulukhaq realized the danger they were in and cried: "The ice is broken."

They started running towards the shore but it was too late: they were already drifting westward along with the ice. They built a snow house and the following day, by the time the ice had stopped drifting, Nulialik urged them to try again to reach the land. However, they did not succeed and had to come back to their *iglu* where they remained for a good part of the winter. They were lucky enough to have among them real shamans who saved them from disaster by preventing the ice they were on to be crushed by an iceberg and by performing the rites that would bring them a good wind.

One by one, they let their knives sink in the water and offered them to the spirit of the sea. The last knife to be offered was a weapon made of solid raw copper; it floated awhile before sinking.

Qorvik also took a small block of ice and threw it towards the land. At the same time he asked the spirits to return them home, safe and sound.

A short while afterwards the wind changed direction and brought them back home. The men leapt from an ice block to another, finally making their way back to the shore. They reached it by the time darkness was falling.

They yelled with joy, ate snow, cried, laughed, and walked home where they found their wives. Some of them, thinking that they were dead, had taken other husbands.

Qingalorqana felt for a long time as though the roll of the ocean was still in his body, waking him up during the night.

—Recorded in the Coppermine region, 1958

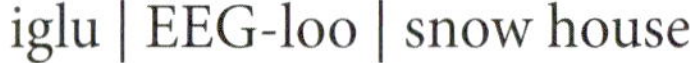

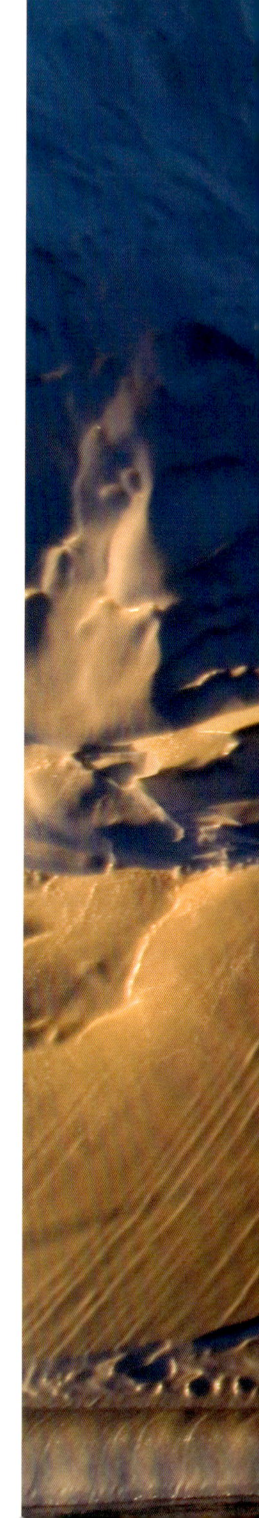

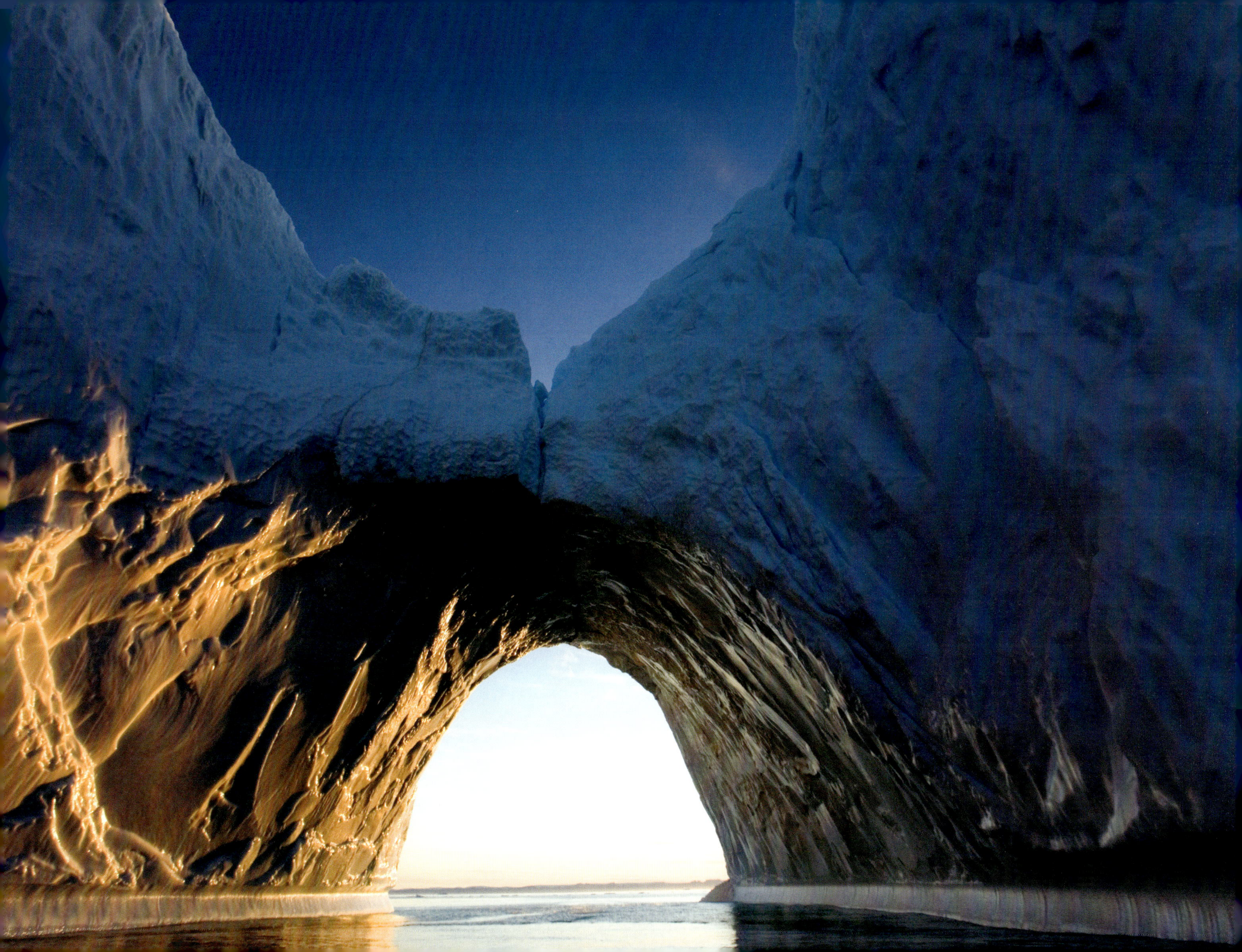

During the dark nights of midwinter seals are netted beneath the ice. The blowholes of the seals are located during the day; at night the hunters go out and make four holes in the ice, in the form of a square, at equal distances from the seal hole; a square net is then placed under the ice by means of a long pole and a cord, so arranged as to cover the access to the hole from below, and held in place by cords passing up through the holes in the ice. When the seal rises to breathe it becomes entangled in the net and is captured. This method of netting is common from Bering Strait to Point Barrow.

Another method of netting seals through the ice was observed on the shore between Bering Strait and St. Michael. In swimming along the shore the seals are obliged to pass near the rocky points and headlands. Taking advantage of this, the hunters make a series of holes through the ice at intervals of from 10 to 15 feet, and then, by use of a pole a little longer than the distance between the holes, a stout sealskin line is passed along from hole to hole until the cord is run out to the distance desired, and is used to drag the long net below the ice. Sinkers are fastened to the lower edge of the net, and it is held in position at each end by a stout cord tied to a crossbar at the hole or to a stake set in the ice. While swimming beneath the ice during the night the seals become entangled in the net and drown.

—*Edward William Nelson, recorded in the Bering Strait, 1901*

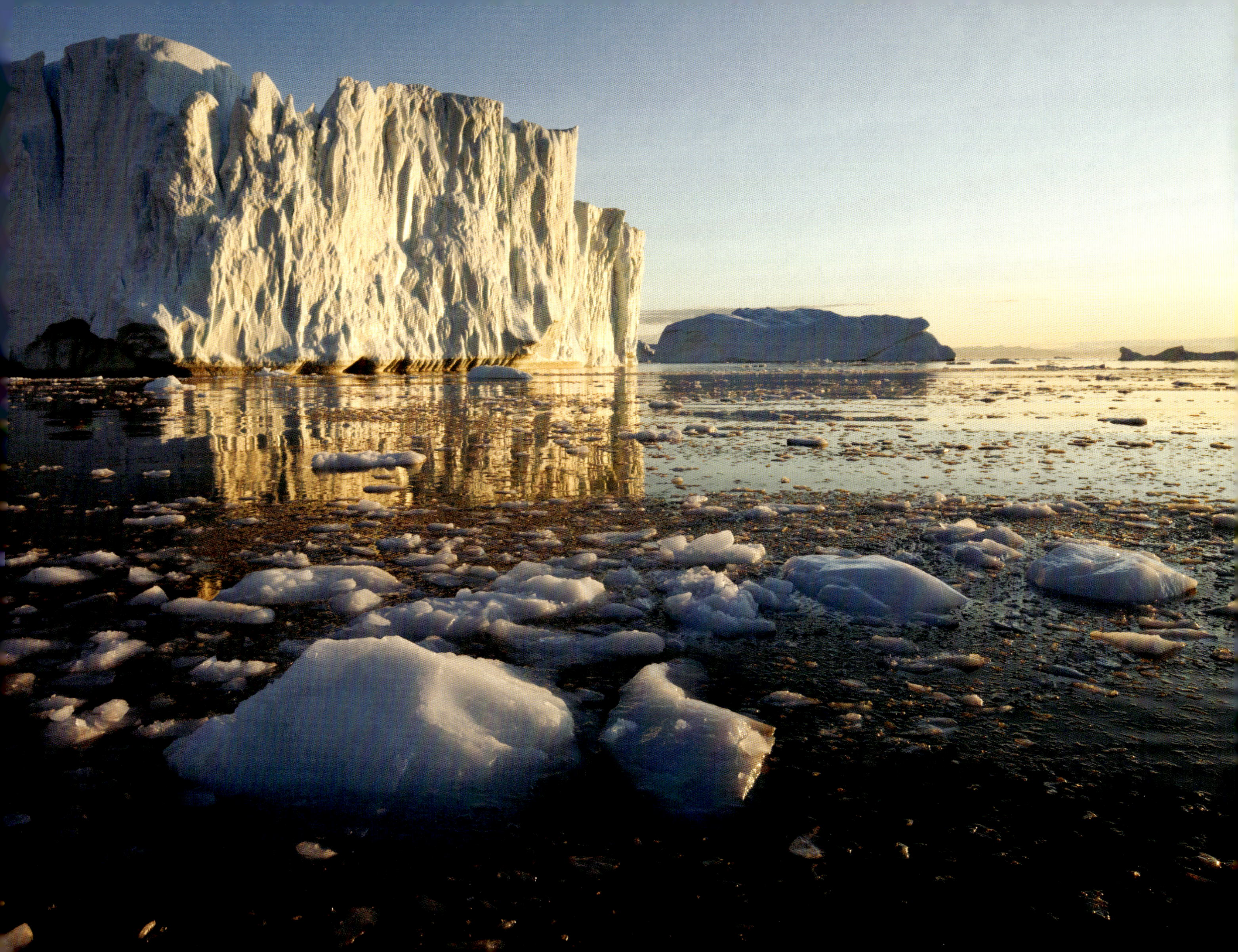

The ice has changed quite a bit from how it was when I was younger, and it is still changing. The ice today seems to be softer and not as hard as it used to be. It is now starting to melt sooner and faster from what I remember as a young lad. The sea ice, whether it is shallow or deep, is starting to melt quickly. Icebergs would stay a while before melting, but today, they melt sooner. Our younger generation must be taught about safety travelling on the ice. There is a difference between thin and thick ice. When ice is thin, it is darker and has frost on the surface because it is just forming or freezing. . . . Even when the sea ice is as thin as about half an inch, it is safe to walk on. Ice is tested with a harpoon by hitting it. When it is hit once, and it breaks, then it is dangerous. . . . The water farther down [from shore] does not freeze at the same time as the water closer to land. Thin ice forms down at the floe edge, so from the wind and the currents of the water, it layers up from the moving waves, and that's how it gets thick [close to land].

—Dominic Pingushat, Arviat

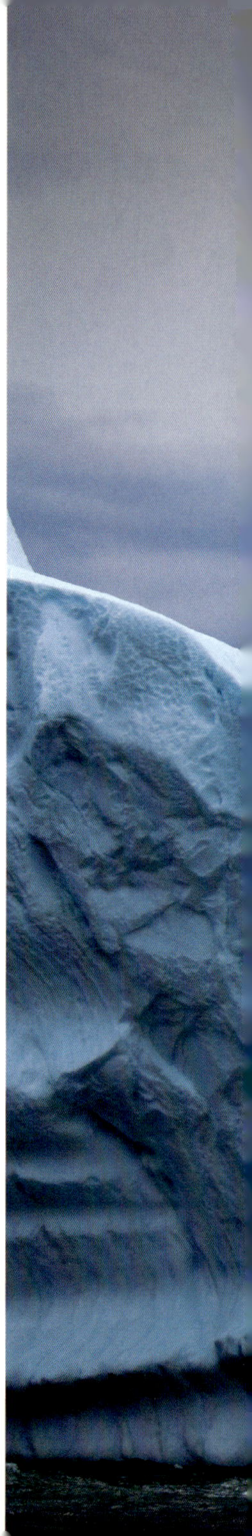

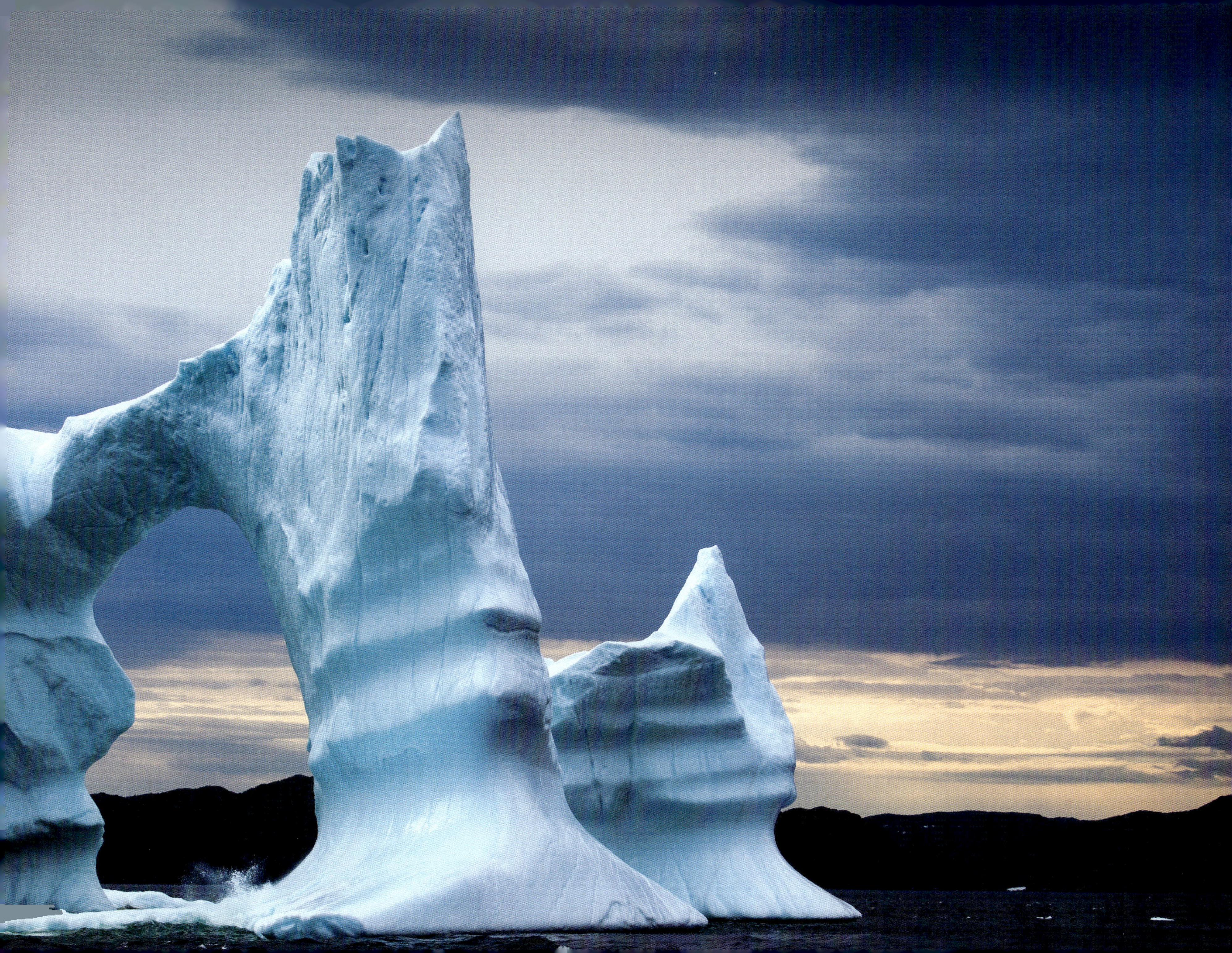

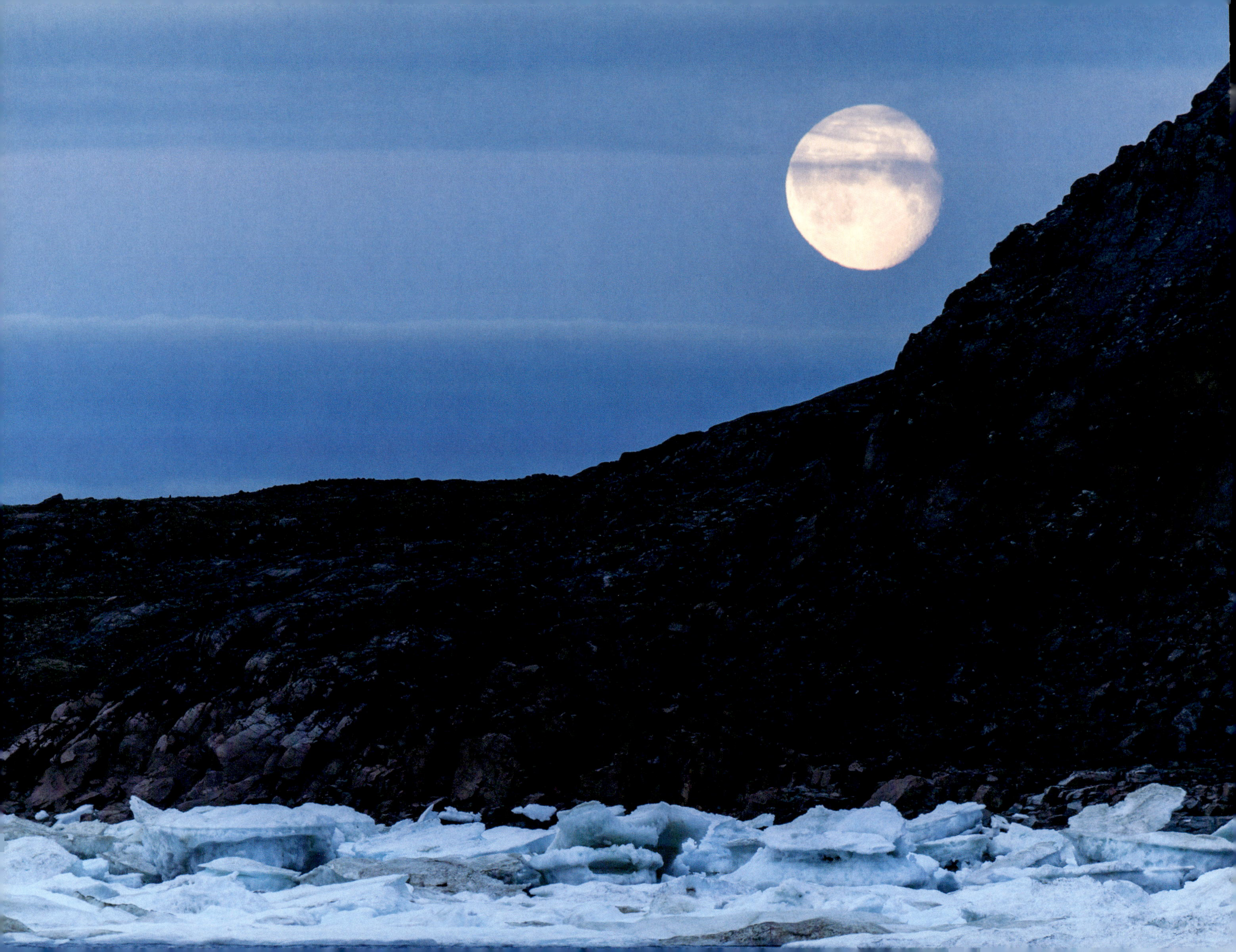

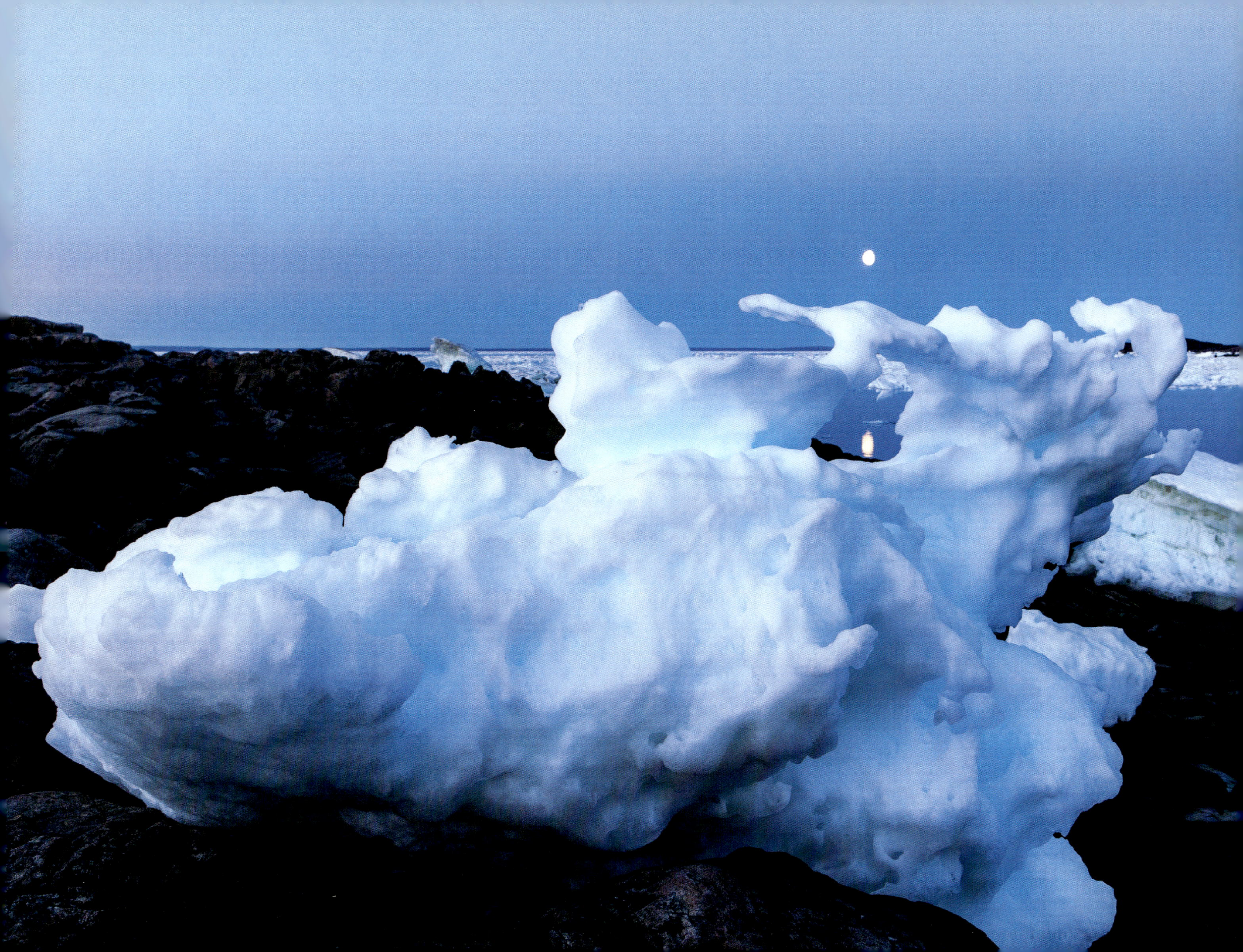

If the ice is really smooth, then it is dangerous when you are down at the sea ice. You should not start walking on the ice without checking it. Always check it with your harpoon first before you go ahead. The harpoon can save your life. . . . When the wind is calm, and the sea ice has formed thick enough to hold a man's weight, it can be safe but must be tested with a harpoon. If it has crystals, glittering spots, then that is an indication that it is thin ice. Inland, for freshwater ice, we were told to watch how the ice looks. If there are cracks on the ice, it is safe to cross, but if there are no cracks, don't cross it. The ice that is just forming has no cracks. Ice that has cracks seems more dangerous, and many people get scared to cross it. If you are going on ice between two pieces of land, it can be thinner than the rest of the lake because a current is flowing, and it is usually thin throughout the winter.

—*Tommy Ublureak, Arviat*

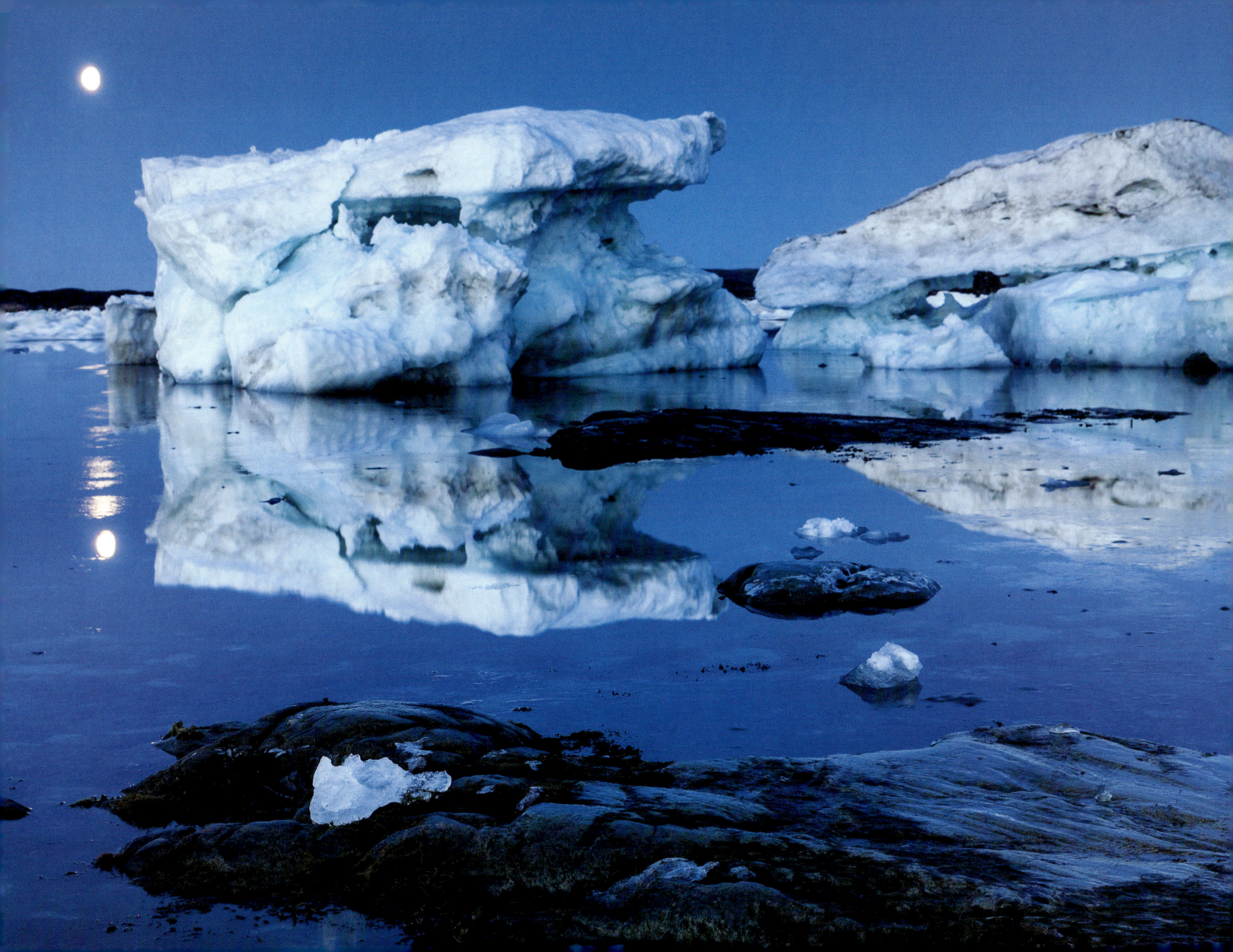

About the end of February the Inuit from Bering Strait southward begin to hunt seals at the outer edge of the shore ice, where the leads are open at that season. On the 28th of February, 1880, I met a party of people on their way from the head of Norton Bay to Cape Darby, where they were going to hunt seals on the ice until spring opened.

At midnight on March 28, the same season, I reached a village on the northern shore of Norton Sound as a party of seal hunters came in from the outer edge of the ice, bringing several fine, large hair seals. . . .

At this season, also, the people about St. Michael begin their usual spring hunting upon the ice. They leave their village, hauling their kayaks, spears, guns, and other implements on small, light sledges made specially for the purpose. Whenever open water is to be crossed the kayak is launched, the sled placed upon it, and the hunter paddles to the opposite side, where he resumes his journey upon the ice. The method of obtaining seals at this time is by the hunter concealing himself on the ice close to the water, and from this point of vantage shooting or spearing them as they swim along the edge. Sometimes a seal is shot or speared while lying asleep on the ice.

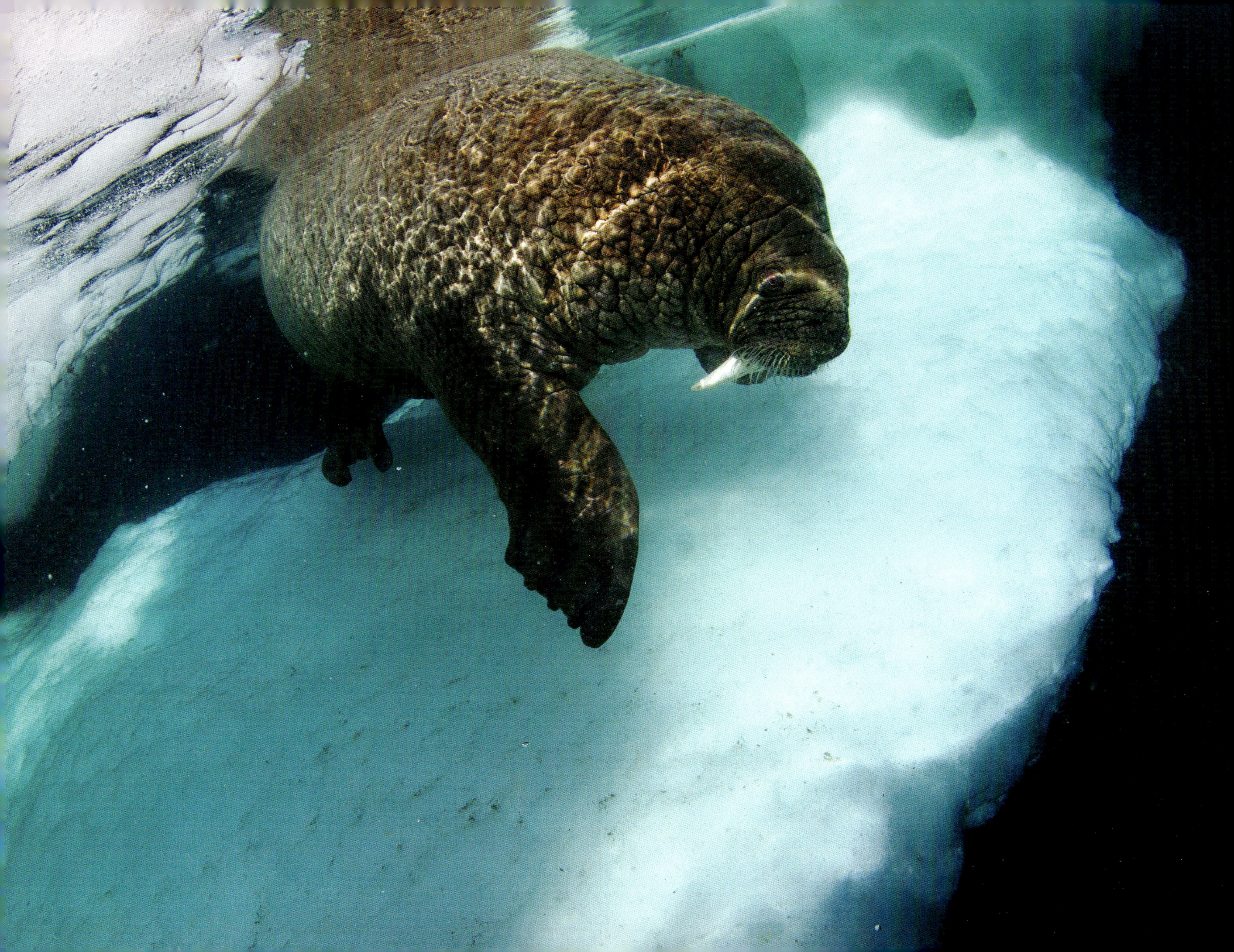

When the ice breaks up, so that there is much open water, with scattered floes and cakes of varying size, the hunters make long hunts in their kayaks, searching for places where the seals have hauled up onto the ice.

On the 10th of May, one season, I met a party of Inuit between Pastolik, near the Yukon mouth, and St. Michael. They had *umiat* of ordinary size on sleds, drawn by dogs, and were going with their families to the outer edge of St. Michael Island to hunt seals, planning to return to the Yukon mouth in the umiat when the ice had left the coast.

During the early spring months the small hair seals come up through holes in the ice to be delivered of their young. These holes are sometimes covered by the hunter with an arch of snow, and the seals are surprised and speared as they come up. . . . Armed with a spear, which has a long line fastened to a detachable point, the hunter approaches erect as near to the seal as is prudent, then lies flat upon the ice. . . . When stalking a seal in this manner the hunter carries a small wooden scratcher, consisting of a neatly carved handle, tipped with seal claws. If the seal becomes uneasy or suspicious, the hunter pauses, and with this implement scratches the snow or ice in the same manner and with the same force as a seal while digging a hole in the ice. Hearing this the seal seems satisfied and drops asleep again. This is repeated, if necessary, until the hunter is within reach of the animal, when he drives his spear into it, braces himself, and holds fast to the line. If close to a hole, the seal struggles into it. By holding the line the hunter prevents its escape, and the animal soon drowns and is hauled out.

—Edward William Nelson, recorded in the Bering Strait, 1901

umiat | OO-mee-aht | skin boats

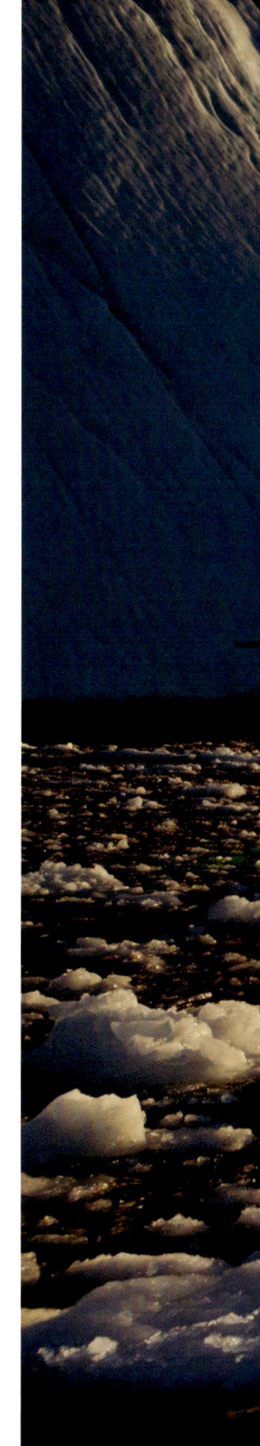

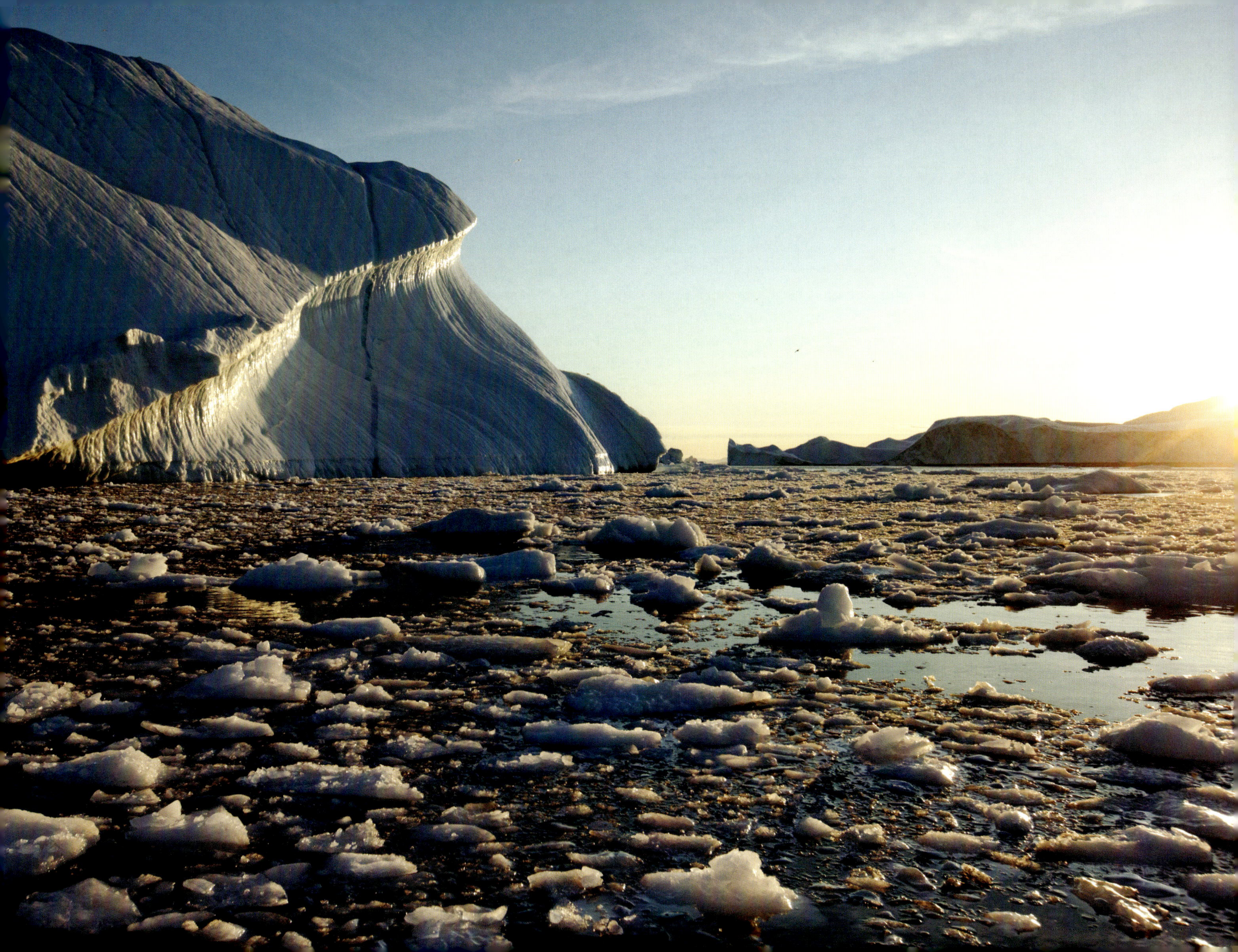

Many communities in the North depended on ice to travel by land and sea. People waited for the ice to form so they could start travelling. There is an Inuktitut calendar in which moons are used to track time (rather than months). One of the moons is called *Tusaqtuut* (November), which roughly translates to "it's time to hear the news." It was a long-awaited month, because it meant the ice would freeze and people could start travelling to other communities to give and receive news. When the ice melted in the spring, communication with other communities stopped because the highway no longer existed. People could not travel by boat very fast, and besides, they had summer harvesting to do for the winter.

Ice is still used for travel today. It is not only the sea ice, but also frozen lakes that create the pathway. When the ice is formed and safe enough for travelling, it becomes a highway without a border. The highway without a border means there is no speed limit, and you can go anywhere anytime, as long as the pathway is safe. Sometimes travel is impossible: in some areas, crushed ice jammed together can create mountains of packed ice too rough for travelling, or a polynya (open water caused by a current) or thin ice caused by a current can form. Usually, polynyas and thin ice are found in the same place every year—places where there's a fast current at a narrow path or land point.

The ice highway makes it easier to travel by land and sea. People in the North travel on ice to go to other communities to visit. The ice is also used to harvest country food; many types of animals are harvested during the winter months using the ice highway.

—Solomon Awa, Iqaluit

Tusaqtuut | too-SAHK-toot | November / it's time to hear the news

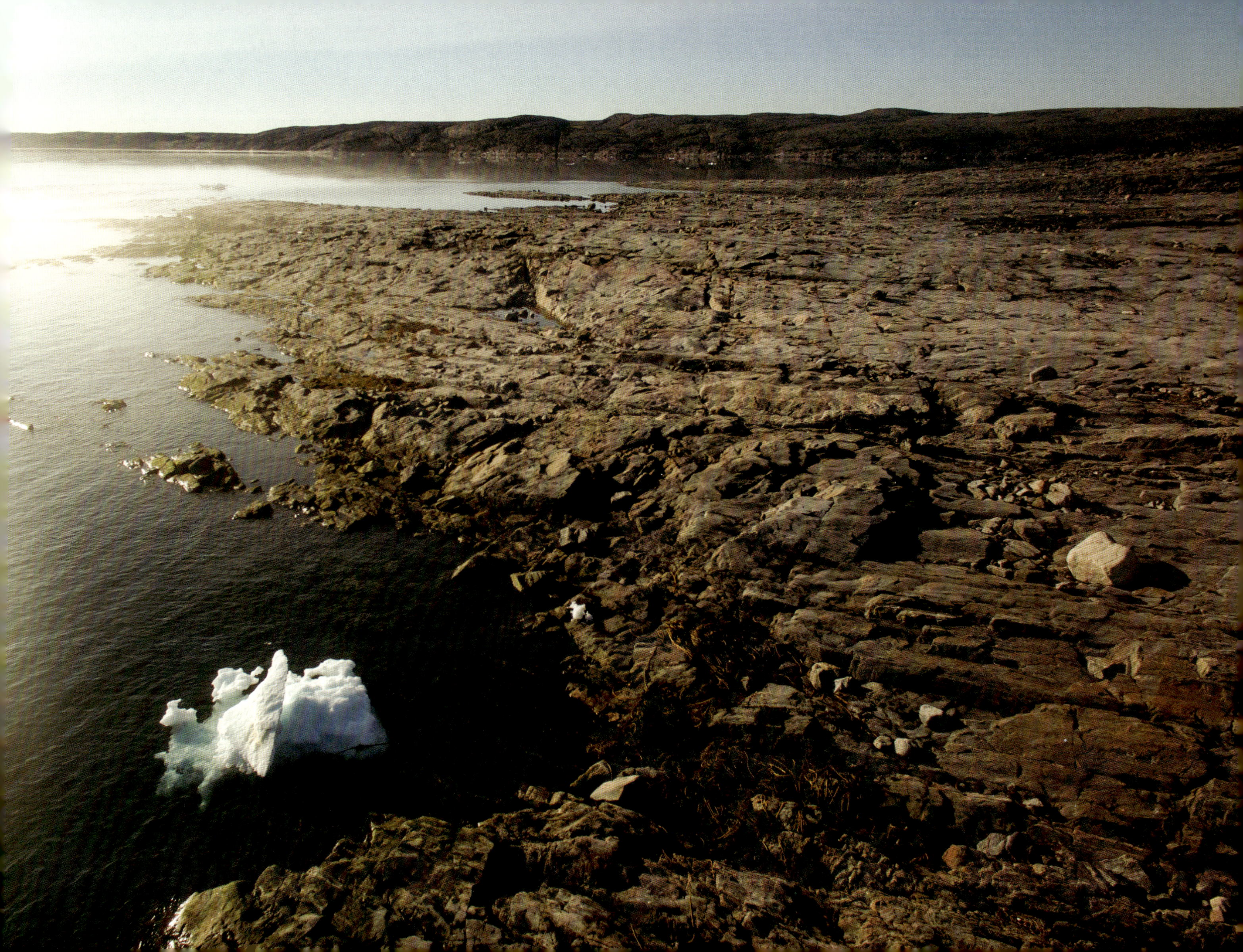

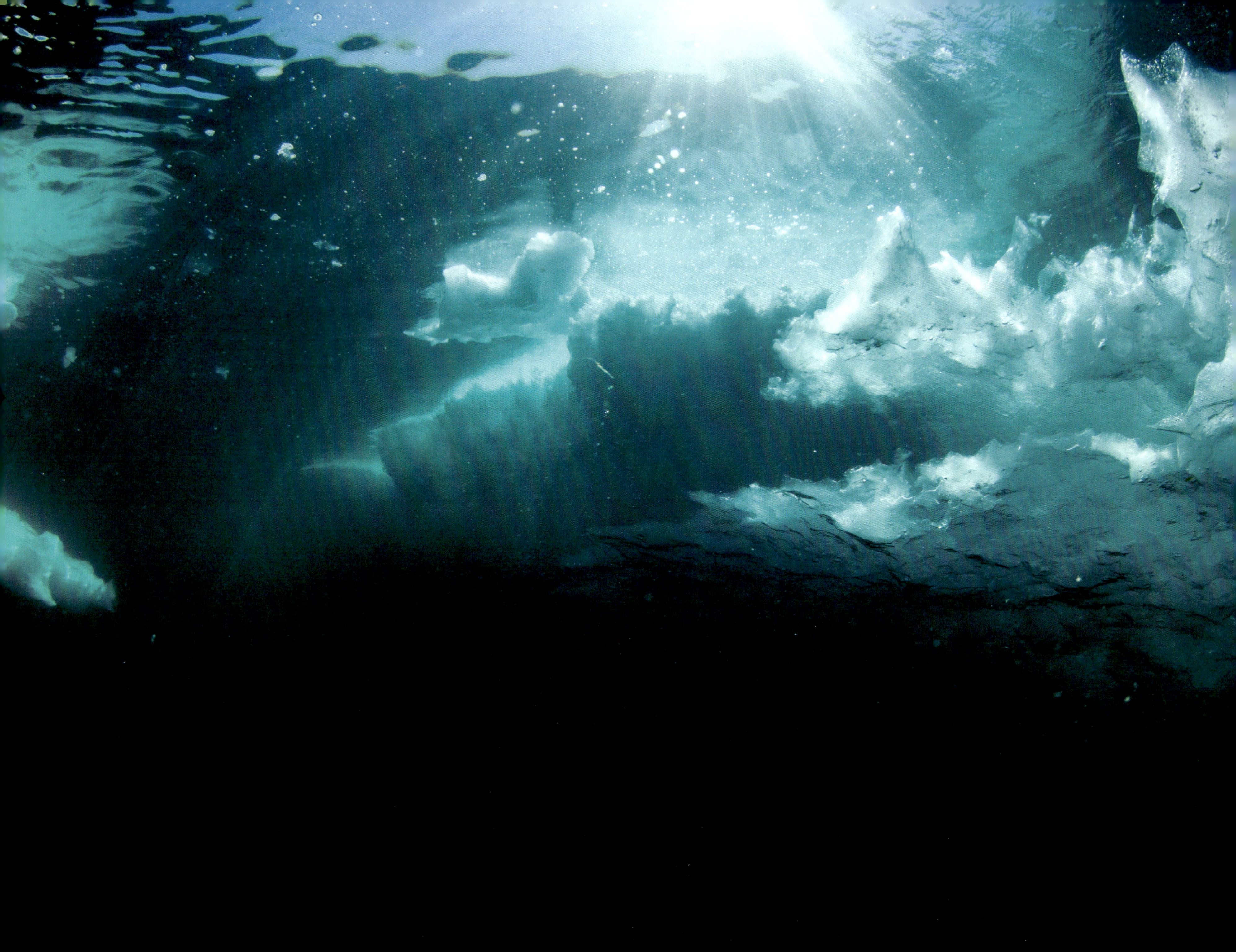

An Inuit tradition that we were encouraged to follow is when a man goes hunting out to the sea, he must not be angry or upset with his wife. Life should be taken very seriously. He should not leave if his wife is not happy with him. It was believed that animals could sense the uneasiness of the hunter and something terrible could happen. My father witnessed that more than once. Hunting was taken very seriously, and animals are not to be joked about, especially large animals and sea mammals. My father once told us about a man who got mad because he did not get a walrus tusk. The men he was with left him, and he went on his kayak and was attacked from behind by a walrus. And something about fish that I know—we now have fishing derbies and try to catch the biggest fish. A family went fishing but did not catch a big one. As they were getting ready to go home, the man caught a fish, but it was too small, so he opened it up and returned it to the hole and said, "You are too small," and the hole became very bloody. Time passed and spring came. The same man went fishing at the river, but his boat capsized, and he went under the ice and was never found. How we treat and what we say about animals is serious, so we must watch how we treat them.

—Tommy Ublureak, Arviat

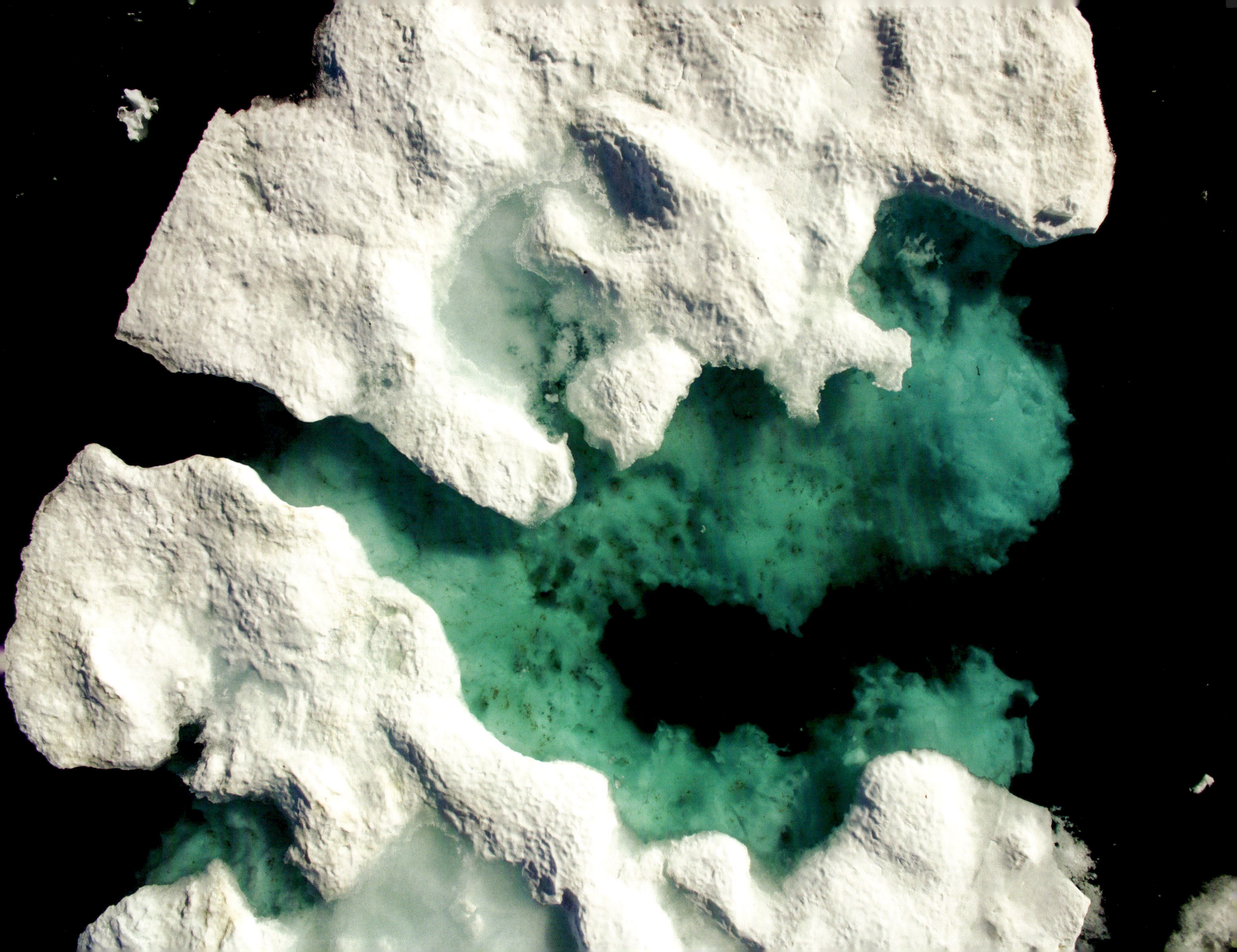

My life's work with signalling to the world the effects of climate change has relied heavily on the observations of our hunters in all the regions and countries where we Inuit live. However, my own observations include how, in 2000, when I moved to Iqaluit, Nunavut, the ice was forming solidly by the end of October. Fast-forward to when I left Iqaluit in 2017: our hunters and families were still boating in December!

In Kuujjuaq, my birthplace, where I returned in 2017, the river that typically was safe to travel across to our trout fishing grounds by November at the latest did not in recent years freeze solidly until January!

In 2021, part of the river that I had a view of from my house did not freeze over at all, therefore making travel more complicated, having to taking a longer route up the river to safely cross.

Our ice in the Arctic is about mobility and transportation, and as the ice becomes unpredictable, it becomes an issue of safety and security first and foremost.

—Sheila Watt-Cloutier, Kuujjuaq

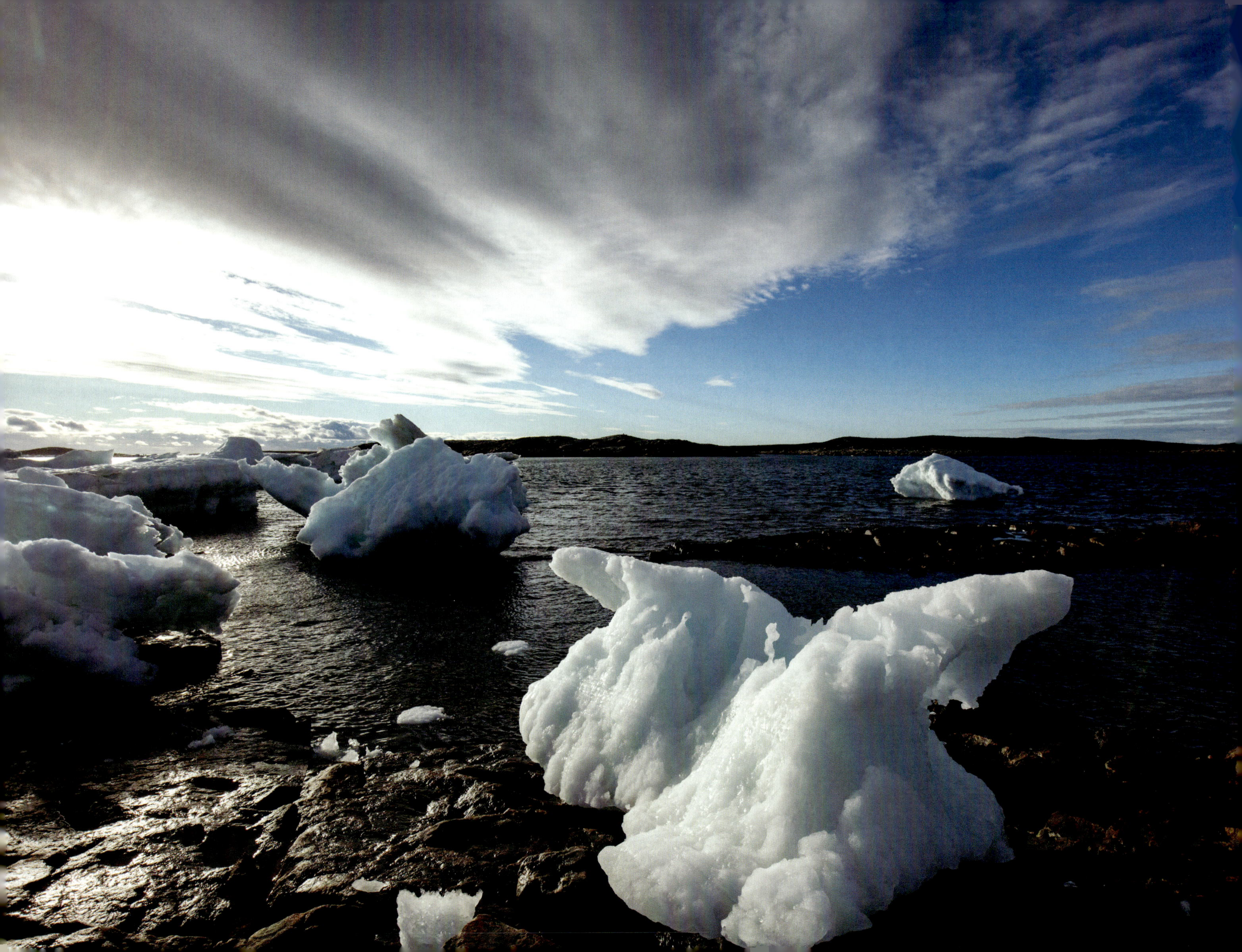

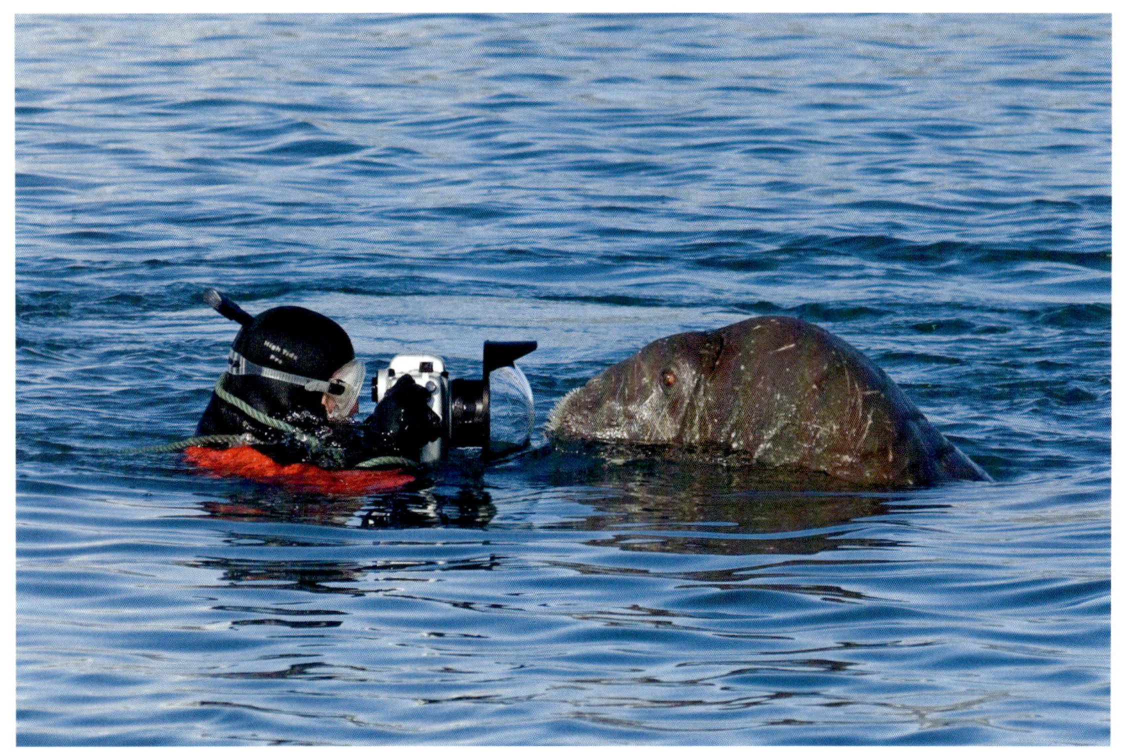

PAUL SOUDERS

Paul Souders is a professional photographer who has been travelling around the world, across all seven continents, for more than thirty years. His images have appeared around the globe in a wide variety of publications, including *National Geographic*, *Geo* in France and Germany, and *Time* and *Life* magazines, as well as hundreds of publishing and advertising projects. His recent photography work in the Arctic has drawn wide acclaim, including first-place awards at the BBC Wildlife Photographer of the Year competition in 2011 and 2013, the *National Geographic* Photo of the Year contest in 2013, the Grand Prize in the 2014 Big Picture Competition, and the Washington State Book Award for *Arctic Solitaire* in 2019. Over the last three decades he has visited more than sixty-five countries and has been slapped by penguins, head-butted by walruses, terrorized by lions, and menaced by vertebrates large and small.

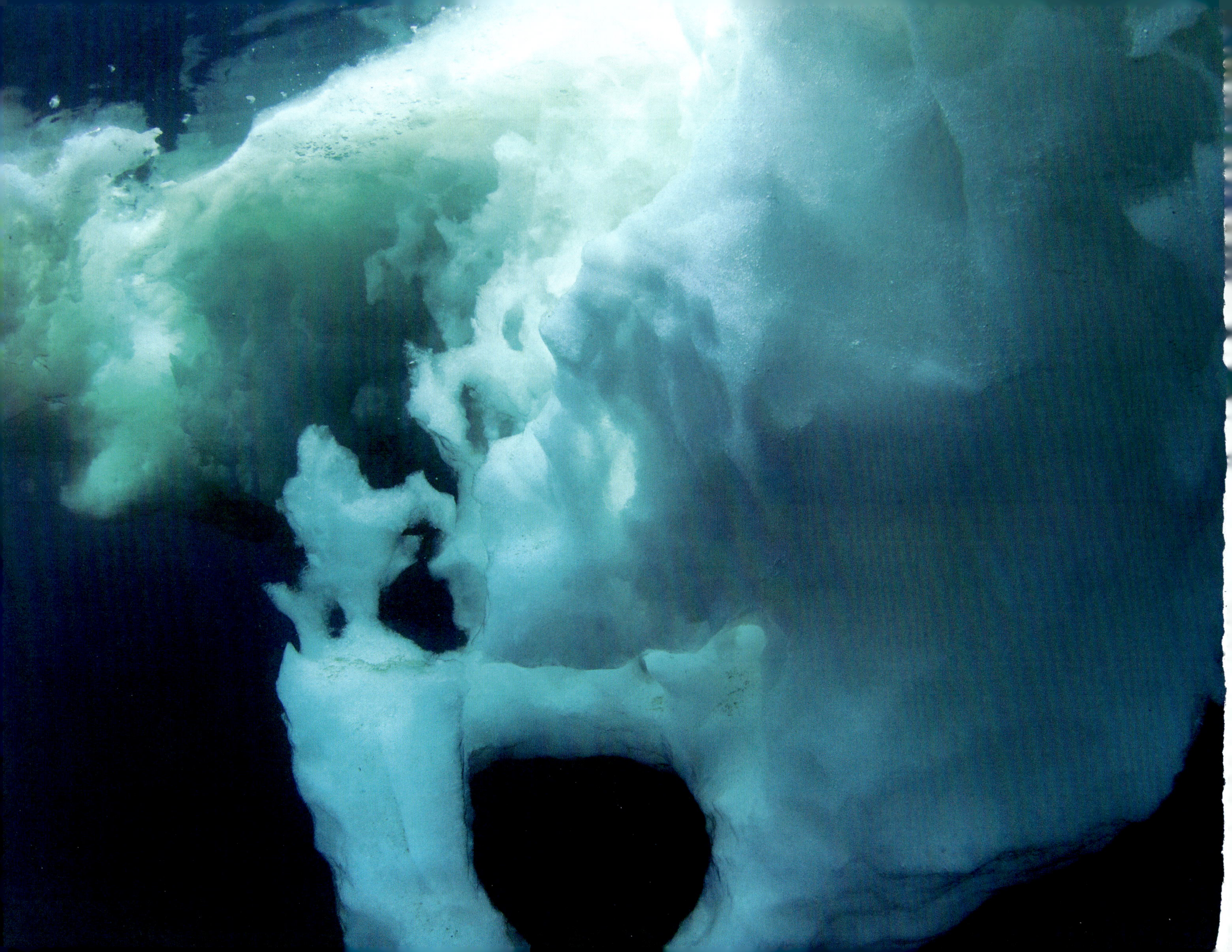